QUEST FOR THE PERFECT STRAWBERRY

Picture 1.
Victor Voth, left, and Dr. Royce Bringhurst, right.
Research scientists and pomologists at University of California, Davis, whose development of many new varieties, growing technologies and techniques revolutionized the California strawberry industry. Unselfish with their time and energy, they have been friends and mentors to many strawberry growers.

QUEST FOR THE PERFECT STRAWBERRY

A Case Study of the California Strawberry
Commission and the Strawberry Industry:
A Descriptive Model for Marketing Order Evaluation

Herbert Baum

iUniverse, Inc.
New York Lincoln Shanghai

Quest for the Perfect Strawberry
A Case Study of the California Strawberry Commission and the Strawberry
Industry: A Descriptive Model for Marketing Order Evaluation

Copyright © 2005 by Herbert E Baum

iUniverse books may be ordered through booksellers or by contacting:

iUniverse
2021 Pine Lake Road, Suite 100
Lincoln, NE 68512
www.iuniverse.com
1-800-Authors (1-800-288-4677)

ISBN-13: 978-0-595-37708-4 (pbk)
ISBN-13: 978-0-595-67547-0 (cloth)
ISBN-13: 978-0-595-82088-7 (ebk)
ISBN-10: 0-595-37708-4 (pbk)
ISBN-10: 0-595-67547-6 (cloth)
ISBN-10: 0-595-82088-3 (ebk)

Printed in the United States of America

To my wife, Gloria, my first editor and critic, whose all around support, help, intellectual input and ability have made the Quest a reality.

The ideas of economists and political philosophers, both when they are right and when they are wrong, are more powerful than is commonly understood. Indeed the world is ruled by little else. I am sure that the power of vested interests is vastly exaggerated compared with the gradual encroachment of ideas. Soon or late, it is ideas, not vested interests, which are dangerous for good or evil.

Lord Keynes

Contents

List of Tables

List of Illustrations

Preface

Three threads finally connected to make this book possible: My involvement in the family's produce business in the 1930s, my early academic background at the University of Chicago, and finally my experiences in the roles of Vice President of Sales and CEO at Naturipe Berry Growers, the largest strawberry cooperative in California from 1958 to 1991, and twice elected board chairman of the California Strawberry Advisory Board, 1975–1976 and 1988–1989. The academic thread is represented by Drs. Theodore Shultz and Oswald Brownlee of Chicago for the agricultural portion and Dr. Milton Friedman for my intense passion and appreciation of the free market. Drs. Milton Friedman, D. Gale Johnson and George Tolley encouraged the writing of this book, as did many colleagues and friends in the strawberry industry and of course my wife and friend, Gloria. Her unfailing support and work over long days, weeks, months, and years have made this book possible.

The industry is rich with important contributions from many of the past CSC board chairmen, including; Ed Zeller, Ray Rose, Tony Ruso, Joe Tomasello, Bill Crowley, Frank Oliver, Bill Deardorff, George Kawanami, Steve Manassero, Miles Reiter, Mitts Nita, George Murai, Ken Morena, Ed Kelly, Bill Ito, George Yamamoto, Bill Moncovich, Mark Murai and Richard Amirsehhi. Malcolm Douglas, Bob Kavet, Dave Riggs, and interim president Cindy Jewell, the only board presidents from 1955 until 2002, were responsible for much of the history in this book. Dave Riggs possessed the unique ability to harness disparate industry people, their energy, experience and intelligence for the common good, and bring the industry into its dynamic research and marketing phase. Other important industry leaders were Frank Heggblade and Joe Margules, founders of Heggblade and Margules Inc., who pioneered forced-air precooling, strawberry and fruit marketing as it is known today, and cooperated with Walton Crane and Harry Long to introduce the first cardboard carton and plastic pint containers.

It must be recognized that the *Quest for the Perfect Strawberry* began with Dr. Harold Thomas, Earl Goldsmith and Driscoll Associates, followed and advanced by the rest of the industry with the previously mentioned leadership as Victor

Voth and Dr. Royce Bringhurst carried the banner. A new generation of players led by Drs. Douglas V. Shaw and Kirk D. Larson, pomologists at University of California, Davis (UCD), has continued the quest, which is perhaps unreachable. However, numerous other private breeders are also seeking the perfect strawberry and are encroaching on the UCD pomology program and reducing its share of the industry acreage. Moreover, there has been a simultaneous deterioration in the past partnership between the California Strawberry Commission (CSC) and the UCD breeders. The apparent lack of teamwork between the other UC researchers and Drs. Shaw and Larson, including prohibition from test stations, has divided the research segment of the strawberry industry, reduced overall research effectiveness and placed in jeopardy the past 60 years of cooperation between the industry and university researchers.

Jimmy Nakamura, of Oxnard, Kunita Shinta, Larry Galper from Watsonville, and Ed Pozzi from Salinas were major grower cooperators with Victor Voth and Dr. Royce Bringhurst, providing their expertise, enthusiasm, and interest in expanding the horizons of strawberry pomology and horticulture. Without their contribution of land, agricultural equipment, labor essential to commercial preparation, planting, and harvesting, plus supervision and observation, the strawberry revolution may not have taken place as rapidly as it did. Shinta and Galper were chiefly responsible for the development of the UCD Watsonville test station. Many other growers have provided time and expertise to make all these forward strides possible, especially Fred Mukai of Watsonville and Joe Kito of Santa Maria. At present, the university breeding program has found other cooperators from the next generation of strawberry growers. Growers such as Thom Flewell, Rod Koda, Darryl Uyeda, and Daren Gee are working together with UCD strawberry breeders Shaw and Larson to continue in the search for the perfect strawberry. The Loftus family, Clarence Pope-Joy, Ralph Wheeler, the Sakuma family of Norcal Nursery, and finally, Bob Parker, Ken Elwood, and Roger Hamamura of Lassen Canyon Nursery were the major forces in conjunction with the Voth-Bringhurst effort to develop new varieties and the horticulture practices necessary for outstanding nursery quality and production. For the past 35 years Roger Hamamura has been my mentor as a dedicated Voth advocate for conveying knowledge and understanding of the varietal-horticultural aspects of strawberries. Voth and Hamamura carried the university banner to all growers, friend or foe. The potential disintegration of the 60-year cooperative research effort between UCD, the CSC, growers, and nurserymen, has reached a critical juncture. Not only has the UCD breeding program fallen behind, but also the cooperation between other scientists and the university breeders has reached a new low. This cooperation, once the ideal partnership between academia and an industry, which previously enhanced both an industry and a large public university in equal degrees, has apparently become a matter of individual ambition overriding the general goals of a common commodity. Industry leaders must make some important changes that will determine the future of the entire non-proprietary industry; the CSC is the only vehicle for making these needed changes which are outlined in this book and include a reduction in the mandatory assessment rate, more emphasis on pomology and horticulture, and less, if any, public relations activities. The September, 2005 departure of the CSC president has provided the perfect moment to evaluate the future CSC mission and the ability to accomplish its goals with appropriate barometers for measuring policy effectiveness. Grower accountability is essential and barometers of CSC policies should show progress or be discarded, new measurable policies begun, and/or assessments reduced.

I am indebted to Victor Voth, Larry Galper, Roger Hamamura, Tom Am Rhein, Curt Gaines, Louis Ivanovich, and Dr. George Tolley for reading my manuscript and making valuable suggestions. Cindy Jewell, Craig Moriyama, Sam Gabriel, George Faxon, Mat Kawamura, Bob Rieneke, Steve Yamamoto, Mark Yatsuya, Miles Reiter, Paul McHaney, Mike Hollister, P.J. Mecozzi, Dave Riggs, Gary Wishnatzki, Ira Nathel, Shelly Nathel, and Stuart Gilfenbain have contributed invaluable information about the strawberry industry. Drs Chalfant, Carter, and Goodhue, Frank Han, Tian Xia, UC Davis, Department of Agricultural and Resource Economics have all provided special insights into the economics of the strawberry market. The consumer research of Dr. Dale Achabal of Santa Clara University helped me to understand changes in behavior and attitudes among strawberry users. I would like to express particular appreciation to my friend Tom Paine for introducing me to the initial stages of the Southern California berry deal in 1955. Tad Tomita, George Kawanami, Hiroshi Shikuma, Tom Iwanaga, and Ed Pozzi had the foresight and courage to encourage the Naturipe Berry Growers cooperative to expand into all areas of strawberry growing and technology which enabled me to gain the knowledge and experience necessary to complete this book and enjoy 50 years in a pioneering and revolutionary industry.

Finally, we acknowledge the valuable contributions of Japanese Americans, from share croppers to entrepreneur growers, shippers, and processors and Mexican Americans who recently have become the new generation in strawberry agriculture that has similarly moved up the ladder in America's pluralistic society, making up more than 56% of the state's 518 strawberry growers compared with 14% of Japanese Americans. Some of their pictures are presented here.

Introduction

This case study of the California Strawberry Industry from 1945 until 2004 is an historical update of a previous book, *History of the Strawberry From Ancient Gardens to Modern Markets*, published in 1974. In addition, this study presents a descriptive *model* for evaluating the agricultural research and marketing aspects of the California Strawberry Commission (CSC) and provides a blueprint for analyzing all federal and state marketing orders (See appendix). The blueprint for analysis is necessary for all marketing boards in order to be accountable to growers, much as an annual financial statement.

Beginning in 1945, the development of the *university varieties*, Lassen and Shasta, created a tremendous expansion in acreage, yield, production, and farm value until 1957, when a setback was caused by growers' "irrational exuberance" and plant disease. The rapid industry decline that followed, due to disease, inferior fruit quality, and excess supply, created the economic necessity for a revolution in pomology, horticulture, and marketing. Thus the stage was set for the grower, shipper and processor to create an industry organization or commodity board, the California Strawberry Advisory Board (CSAB), which later became the California Strawberry Commission (CSC), to finance and coordinate expenditures on pomology, horticulture, and marketing, and to establish a plan for the future. Table 1 illustrates this rapid growth, disastrous decline, and continuous industry expansion from the depth of economic despair. All tables provided herein describe the historical record in figures with the "rest of the story" to follow in narrative form.

The California strawberry revolution, illustrated above, resulted in the competitive destruction of all regional strawberry growing areas in the United States and established its world preeminence in strawberry pomology and horticulture.

In addition to the historical perspective and update, this study will evaluate the CSC, utilizing a *CSC-university model* for measuring the effectiveness of its pomological and horticultural expenditures on the supply of strawberries, production per acre and cost of production, total per acre returns, as well as the demand effect of quality improvements from new varieties. The effectiveness of the CSC's budget for marketing (advertising, promotion, public relations, food/health research, etc.) will be evaluated by incorporating those functions into a descriptive model (Rauser) to study the relationship between the demand for fresh strawberries and those variables affecting demand. The central question is whether the CSC's mandatory check-off programs, or assessments, have caused demand to increase and whether that increase has had a positive effect on the welfare of California strawberry growers. The information developed from my descriptive model will be used to understand the strawberry market and to describe the basic features and internal functioning of the system. The model explains the role of the CSC, its relationship to the private sector grower-shipper and the separate and joint measurable effects on growers' welfare. In order to assess the variables that effect demand, there is need for some *tangible asset* (Blaylock) to justify supporting a marketing order, for example, an increase in fresh and freezer price, retail ads, changes in attitudes and behavior, and quality improvement of new varieties, better packaging, and refrigeration. Thus, barometers for evaluating expenditures are essential and helpful in measuring or describing the effect of a program or programs. These barometers will enable the measuring and describing the quantitative or qualitative impact of price changes, promotional-advertising programs, public relations, varietal and horticultural advances and other variables effecting demand for strawberries. In a later chapter, the history of the CSC, and its research on many advertising and promotional programs, will be evaluated using barometers such as number of ads compared to prior years, sales dollar contribution per store, strawberry consumption by light and heavy users, strawberry market share compared to other fruit, and strawberry category contribution to total produce sales dollars. A further example of measuring the effectiveness of CSC's varietal and horticultural expenditures would utilize average nominal and real prices, total sales, production, yields per acre, costs per acre, and total returns per acre, as well as the previously mentioned barometers. This model could be the basis for further study of the CSC and other Boards.

The major problem facing agriculture today is excessive production with one of the remedies being the development of federal and state marketing orders and commodity boards. The relative inability to solve overproduction through marketing orders has led to producers' skepticism regarding the role of commodity boards and marketing orders. This grower reluctance to continue funding apparently ineffective programs (no tangible asset) has progressed into the courts where, as one example, a 2005 Florida Court of Appeals ruled in support of the citrus growers' argument that free speech rights were violated by the *tax* or mandatory assessment. Subsequently, the constitutionality of the state's *box tax* is under review by the Florida Supreme Court. Prior to this, the 2002 United States

Supreme Court ruled that a national mushroom marketing order could not conduct generic advertising on behalf of that industry, because of the violation of the grower's right to free speech. Since the CSC is no longer involved in generic advertising, and emphasizes only research, the free speech question may not arise to jeopardize the CSC mandatory assessment. However, the 2005 Supreme Court decision on the beef marketing order ruled that grower *free speech* was not infringed upon by the mandatory marketing order assessment, because government was considered to be the *speaker*, since this was a federal marketing order. In the future, government may become more involved in marketing order advertising and promotion programs in order to insure that *message attribution* be government sponsored or approved, and therefore not subject to grower challenge. I hope this ruling, although Orwellian, does not infringe upon the independence of state and federal marketing orders.

The following table is a summary of Tables 6 and 7.

Table 1
California Strawberry Production & Value: 1945–2005

Year	Acres	Total Production Pounds	Tons Per Acre	Fresh Pounds	Pounds (+ juice berries)	Freezer Farm Value* Fresh	Farm Value** Freezer
1945	1,100	n/a	n/a	n/a	n/a	n/a	n/a
1950	5,700	81,282	7.1	47,724	33,558	$ 9,545	$ 6,712
1957	20,700	223,560	6.4	118,260	105,300	21,169	10,951
1967	8,000	208,800	13.1	148,034	60,766	34,359	8,923
1975	10,000	378,996	18.9	269,859	109,137	91,835	20,402
1990	20,000	982,904	24.5	666,519	316,385	345,369	85,450
2000	26,300	1,452,046	27.6	1,045,118	421,409	673,995	93,311
2001	25,142	1,267,241	25.2	942,191	338,960	729,731	96,747
2002	27,200	1,473,645	27.3	1,070,796	431,787	852,843	137,740
2003	28,200	1,622,681	28.7	1,182,697	471,455	925,368	126,124
2004	31,600	1,670,463	26.4	1,224,587	476,391	1,050,329	118,711
2005	32,636						

*   Fresh Value based on FOB prices quoted by USDA Fruit & Vegetable Market News (not actual prices after adjustments for poor quality, weakened market, etc.)

**   Freezer Value based on FOB prices quoted by Processing Strawberry Advisor Board

CHAPTER 1

Historical Perspective: An Update 1945–1972

Significant agricultural research occurred at the University of California prior to 1953, preceding the leadership of Dr. Royce Bringhurst, pomologist, and Victor Voth, horticulturist, both of UCD. The formation of the Central California Berry Growers (the forerunner of Naturipe Berry Growers) in 1917, a marketing and research organization consisting of early Naturipe and Driscoll growers, was the forerunner of the California Strawberry Advisory Board, which later became the California Strawberry Commission. However, the birth of the *modern* strawberry industry and its unprecedented growth and success proceeding 1953 was due to the prime varietal and horticultural research and development of Bringhurst and Voth. Their work was completely supported by the newly created California Strawberry Advisory Board consisting of growers, shippers, industry, and university financial and technical contributions. The competitive conflict between the strawberry industry and proprietary interests began in 1945 after Dr. Harold E. Thomas and Earl V. Goldsmith, the major University of California strawberry researchers at that time, resigned from the university to form the non-profit Strawberry Institute of California in Morgan Hill, organized by E.F. (Ned) Driscoll. Before departing they named 5 university varieties, Shasta, Lassen, Donner, Tahoe and Sierra; Shasta and Lassen, which became popular in the US, will be discussed later. They also left 1,200 selections, which were being grown at the Wolfskill Experimental Orchard (Department of Pomology, UC Davis), located at Winters. What remains unknown, perhaps to anyone but The Strawberry Institute, is the quantity of planting and nuclear stock, previously developed at the university, which almost certainly became Strawberry Institute foundation stock for further nursery and commercial growing. An inventory of plant materials was unavailable during the strawberry program's transition period and, therefore, neither the strawberry industry nor the university is aware of the nature and degree of the plant loss.

Dr. Richard Baker was hired by the Pomology Department at the University of California in 1945 and Victor Voth as a Research Technician in February of 1946; both were assigned to the new strawberry research program at UCD.

During the next 7 years the strawberry research program lacked funding. Thanks to the political effort exercised in southern California by Fred Yasakochi, George Nagata, Harold Stokes, Jack Tabata and Tad Munemitsu, a research plan was developed and presented to the state legislature and completely funded in 1952. Victor Voth was promoted to Specialist in 1952 at the USDA station, Torrey Pines, with experimental plots at Fallbrook, San Luis Rey, and Antelope Valley. All research was transferred in 1958 to the South Coast Field Station, Irvine. After 8 years, Dr. Baker left the university in April of 1953 and Dr. Royce Bringhurst was hired in September 1953. Thus began the research to remedy the varietal, horticultural, and financial problems which developed following the initial success of the Thomas and Goldsmith university varieties.

These 5 previously mentioned varieties were bred for the California climate and long distance shipping. Because of World War II and the internment of the Japanese-Americans, there were no farmers for the new varieties and the revolution could not begin until the war was over in late 1945 when old and new farmers began the *strawberry industry transformation* in agriculture. Strawberries were grown in every state in the Union prior to the war with a variety adapted for each climate and market. It was a minor industry until World War II due to sensitivity to disease, high cost of production, susceptibility to decay, and inadequate transportation. California was 3,000 miles from its best potential market, the Eastern states.

This first great expansion, which brought a wide variety of non-farm investors from all over the US, was similar to the gold rush of 1848. Since over 70% of the production went to the freezers, the 1957 price collapse from 16¢ to 8¢ per pound resulted in below cost returns to the grower-investor. Production problems were developing because of the sensitivity of berries to soil and salinity, in addition to the insects that suck the life fluids from plants, prevent flowering, halt fruit set, destroy leaves or carry infectious diseases. *Verticillium wilt*, a soil fungus that can wipe out a crop in a short time, became a major production problem forcing growers to continually move to new ground. Berry plants were often retained for 3–4 years compounding soil and insect problems. As a result of marketing and production problems, acreage declined to 8,000 acres by 1967, although yields per acre increased to 13.1 tons. The university varieties, susceptible to all of these production problems, produced for only a short period of time, March through June, and had many quality problems. The freezer market was the only user for the remainder of the year. While the industry was in decline,

Thomas and Goldsmith were developing *proprietary varieties* for Driscoll, Inc., which provided a competitive advantage with varieties of better size, taste and an extended fresh market through October. The university varieties could not be shipped during the summer and fall months because of quality. Of these proprietary varieties, the Driscoll variety, Z5A, became the dominant, non-university variety, and established them as the premier California grower-shipper, a position they still retain.

This book is written to honor two scientists, Dr Royce Bringhurst and Victor Voth, whom author George Wells in his book, *Garden in the West*, introduced as "heroes of these dark times." Wells called them "…iconoclasts, ruthlessly discarding almost every practice then in vogue and substituting new ones, discarding plants and bringing in new ones, knowing the urgency of new varieties to save an industry decimated by inadequate varieties and subject to a formidable array of viruses, bacteria, insects, and slightest variation in temperature, altitude or hours of daylight. Over 50 years ago growers were about to throw up their hands in final and utter defeat, with grave predictions that the strawberry would nevermore be seen growing as a commercial crop in California."

Wells artfully states the role of Bringhurst and Voth: "At this juncture, science came along to stir the ashes of disaster with a magic wand. The result was one of the most fantastic transformations in the history of agriculture. A plant breeder and a horticultural technician, working smoothly together and generously financed by an industry at the firing wall, halted the downslide almost overnight. Since then, the story of the strawberry in the Golden State has tongues wagging and heads shaking all over the world."

This revolution and industry transformation was hardly mentioned in Wilhelm's book, *A History of the Strawberry*. He did, however, accurately credit Thomas and Goldsmith for the 5 university varieties, pioneering new cultural methods and supporting university research on control of soil-borne diseases by soil fumigation with *methyl bromide* (MeBr) and *chloropicrin* mixtures in the crucial years when the idea was being developed by the university, as well as improved crop harvest, transport and refrigeration systems. Bringhurst, Voth, Mitchell, Paulus, and many other university researchers were not credited with the real pioneering developments in new university varieties (1961—Fresno, 1963—Tioga, 1970—Aiko and Tufts) and horticultural discoveries and applications (sprinkler-drip irrigation, annual planting, summer planting, winter planting, high latitude and elevation nurseries, polyethylene mulch, improved methods of chloropicrin-methyl bromide soil fumigation, 4 row beds with increased plant population, slow release fertilizer, and use of the high gallonage sprayer), all of which were responsible for the 1967 to 2004 explosion in acreage, yields, crop value and industry financial success. The apparent obsession with The Strawberry Institute and the merger with Driscoll Strawberry Associates to form Driscoll, Inc. in 1966 probably contributed to the Wilhelm exclusion of the true scientific and horticultural developments prior to 1974, the date of publication. This is illustrated by Wilhelm's quotation: "The work, courage, creativity and acumen of the Associates, devoted to development of superior strawberries, maintenance of extensive, disease-free nurseries and plant-breeding grounds, production, on enriched soils, of fruit of the highest quality, and shipment of it fresh to many parts of the globe, are examples of excellence for the entire agricultural world." The most glaring historic omission was the failure to recognize the beginning and rapid development of the Southern California berry industry, from San Diego to Santa Maria, perhaps since Wilhelm, Thomas and Goldsmith, being from Northern California, were at the time unaware or uninterested in these *early* producing areas which now represent over 60% of total acreage. Although Driscoll tried experimental plots in Southern California, it was much later before suitable proprietary *short day* types were developed.

Bringhurst and Voth, from the beginning of their joint effort quest for new varieties, selected a model for developing short day types, which emphasized earliness, color and firmness. Fortunately, their selection process also produced varieties versatile enough to be successfully used in numerous environments, including Northern California. They recognized that The Strawberry Institute had a monopoly on early *everbearers* and appreciated the need for a quality improvement over the 5 university varieties developed earlier by Thomas and Goldsmith, as well as developing varieties which would lengthen the season and enable non-Driscoll growers to compete during the summer and fall months. As Bringhurst and Voth developed improved short day types, they continued to breed for *day neutral* varieties while The Strawberry Institute/Driscoll Strawberry Associates searched for their own patented short day type.

Although *A History of the Strawberry from Ancient Gardens to Modern Markets*, chronicled much interesting information, it failed to appreciate the major Revolution, sponsored by non-Driscoll industry leaders, which lead to the development of a breeding, horticultural, and marketing model for the present and future. It was this model, with the California Strawberry Advisory Board (California Strawberry Commission) at the center, which provided the grower funded cooperative development of an industry-university partnership. Beginning earnestly in 1952, this partnership continues to flourish; this history and Victor Voth's *unpublished records* provide the material for this study.

Illustrations on the following pages: Industry research leaders of the early years.

Picture 2. Dr. Royce Bringhurst and Victor Voth

Picture 3. Top left to right: Roger Hamamura, Victor Voth, George Kawanami.
Center: Kenny Elwood. Bottom left to right: Royce Bringhurst, Curt Gaines, Kenny Ellwood.

Picture 4. Top left to right: Kuni Shinta, Royce Bringhurst, George Uesugi, George Kawanami. Center left to right: Kuni Shinta, Tad Tomita, Hiroshi Shikuma. Bottom left to right: Tom Ito, Ed Pozzi.

Picture 5. Top: Victor Voth, Royce Bringhurst, Steve Manassero. Center: Larry Galper, Bill Crowley, Herb Baum, Fred Hirosuna, Howard Hall. Bottom: Norm Welsh, Richard Buchner, Wayne Schrader, Luis Valenzuela, Marvin Snyder, Royce Bringhurst

𝔉𝔦𝔯𝔰𝔱 𝔐𝔢𝔪𝔟𝔢𝔯𝔰' 𝔐𝔢𝔢𝔱𝔦𝔫𝔤
April 9, 1917

CONSENT TO HOLDING FIRST MEMBERS'MEETING
of
CENTRAL CALIFORNIA BERRY GROWERS ASSOCIATION

We, the undersigned, being all of the members
of the CENTRAL CALIFORNIA BERRY GROWERS'ASSOCIATION, viz:

Mark Grimes	R. F. Driscoll
Sumito Fujii	T. Sasao
James Hopkins	T. Kato
O. O. Eaton	K. Skikuma
J. E. Reiter	F. J. Moriyasu

P. S. Ehrlich

do hereby give our written consent to the holding of this,
the first members' meeting of CENTRAL CALIFORNIA BERRY
GROWERS'ASSOCIATION, on Monday, the 9th day of April, 1917,
at the hour of 11:15 o'clock a.m. at 444 Bush Street,
San Francisco, California, and we do hereby certify that
all of the members of the association are at this meeting,
now here present.

IN WITNESS WHEREOF, we have hereunto subscribed
our names this 9th day of April, 1917.

Thanks to our pioneering fathers and the above charter members of the cooperative, we are today (Feb. 10, 1967) holding the 50th annual meeting in San Jose, California.

Central California Berry Growers (Naturipe Berry Growers) first meeting charter,
April 9, 1917

CHAPTER 2

California Strawberry Advisory Board and the Processing Strawberry Advisory Board

California Strawberry Advisory Board (CSAB)/California Strawberry Commission (CSC)

In order to secure funding, the industry approached the California state legislature to establish a mandatory industry law to be included in the California Food and Agricultural Code in 1955. The code provided the California Strawberry Advisory Board (CSAB) with the authority to collect assessments from growers, processors, and shippers on all harvested berries for fresh or freezer markets. In 1995 the commodity board's name was changed to the California Strawberry Commission (CSC). The commission system enabled the industry to become more involved with political issues such as global control of methyl bromide, immigration, and food safety. The objectives of both forms of administration were:

"The establishment of the commission is necessary for the efficient creation and management of a research program to develop improved varieties of strawberries, an integrated approach to control pests and disease common to strawberries, and more efficient cultural practices. The commission is also necessary for the efficient development and management of a national and international advertising program which, combined with a research program, will enhance the competitiveness of the California strawberry industry within the national and international marketplace."

This legislation establishing the CSAB with mandatory industry funding established the model currently used for the cooperative venture between the CSC and the University of California. Budgets and priorities were established, and industry leaders formed committees to plan, identify problems and prioritize expenditures. The industry realized that a joint industry-university effort was necessary for emergency progress in pomology and horticulture and a new approach to marketing must be developed.

Processing Strawberry Advisory Board (PSAB)

This Board was formed by the California state legislature on July 7, 1960, and amended in 1961, at the request of growers and processors who had experienced below cost pricing, a demoralized market, poor quality, and unfair trade practices. The purpose of this order was to authorize and require federal and state grading and inspection of all freezer grade strawberries. The Board required processors to post a grower's schedule of field prices and terms at their receiving stations, and administer unfair trade practices with respect to these postings. Promoting the sale of strawberries for the purpose of maintaining existing markets or creating new and larger markets was also a responsibility of the PSAB. However, the PSAB transferred all frozen promotion activity to the CSC, since growers and processors were already required to pay assessments to the CSC. Processors fund and control all PSAB activities separately from CSC functions and are under the supervision of their own board of directors.

Illustrations on the following pages: Board chairmen, board presidents, board members and others.

Picture 6. Top: Joe Tomasello, Malcom Douglas, Tony Ruso. Center: Frank Oliver, Malcom Douglas. Bottom: Malcom Douglas, Bob Kavet, Dave Riggs.

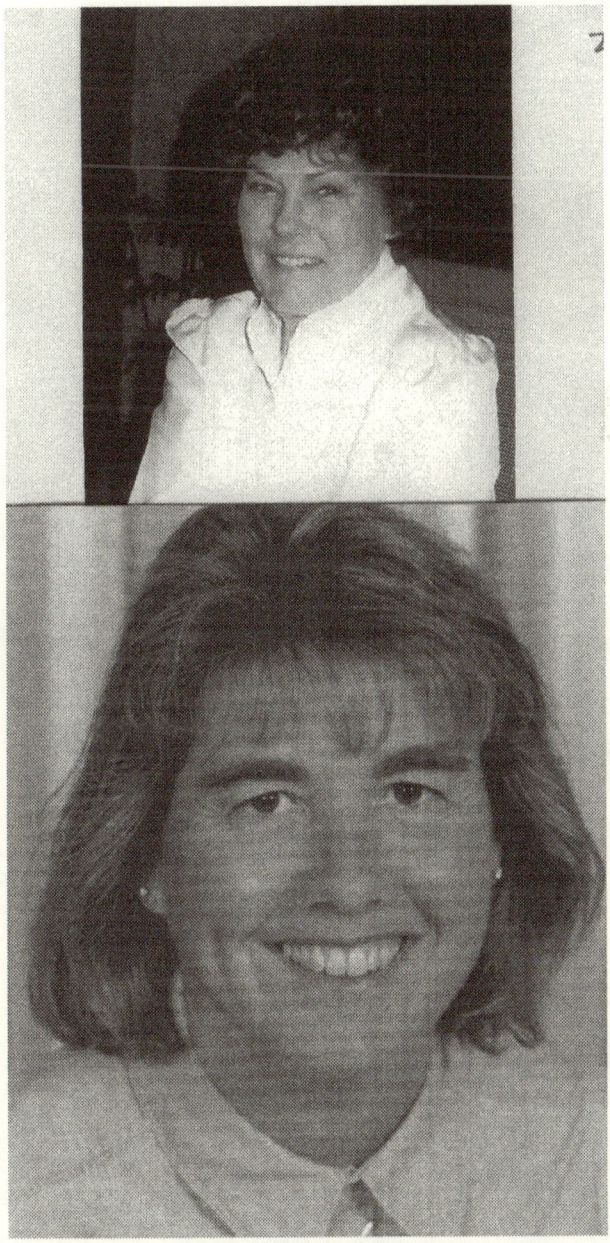

Picture 7. Top: Anabel Ruso, CSC Controller, retired in 1986 after more than 20 years. Cindy Jewell, CSC employee 1/1/1983–10/31/2002, Marketing Coordinator, Marketing Director, Director of Operations, Executive Vice President and interim President.

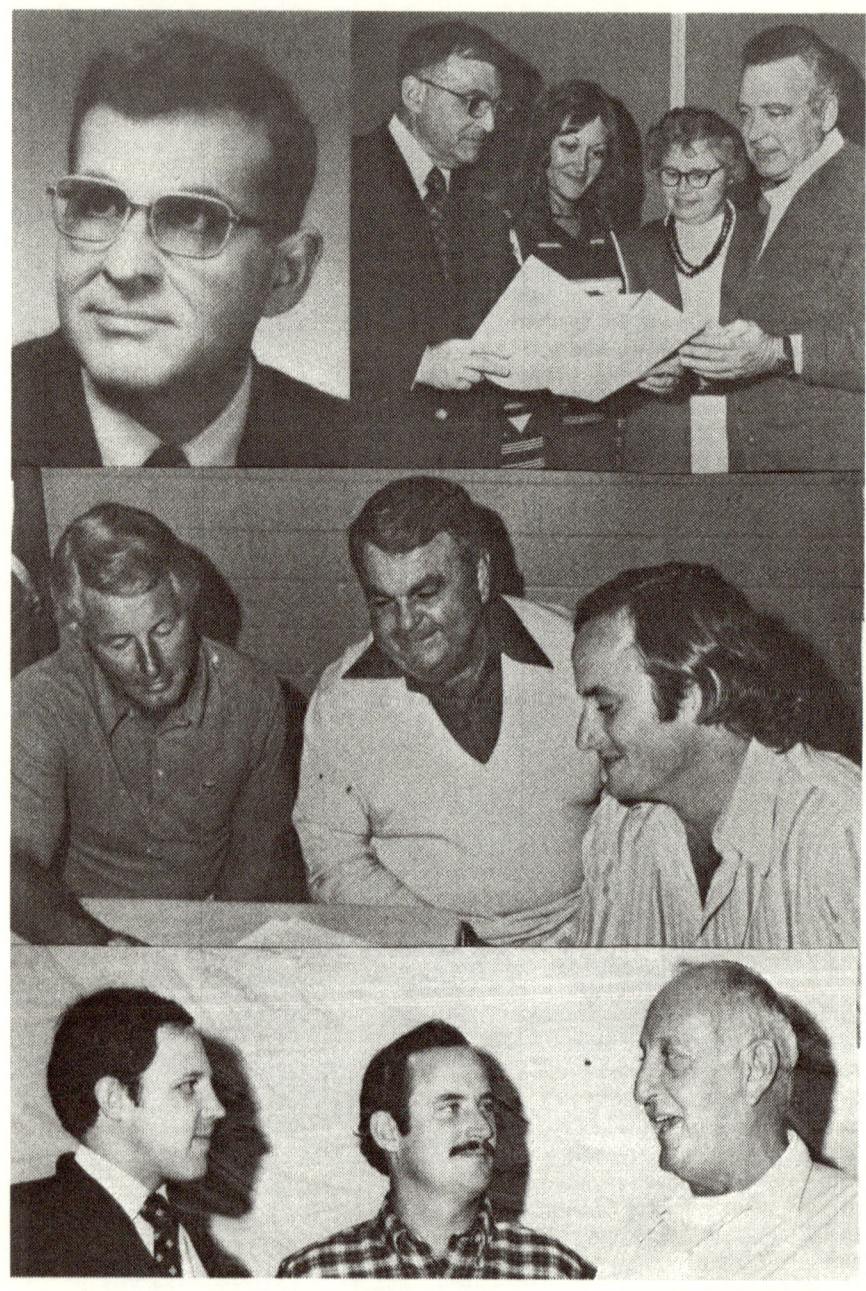

Picture 8. Top: Herb Baum, Claire Biancalana, Jean Downes, Bob Kavet. Center: Bill Deardorff, Bill Crowley, Miles Reiter. Bottom: Dave Riggs, Miles Reiter, Steve Manassero

Picture 9. Top: Mits Nita, Tak Higuichi, Hank Drabin. 2nd Row: George Faxon, Burt Jacobson, PJ Mecozzi 3rd Row: Carlos Raffo, Red Bryan, Mack Ramsey. Bottom: Reid Wagstaff, Doug Circle, Al Yamamoto

CHAPTER 3

Industry Problems and Solutions

Pomology

The history of pomology at the University of California has <u>not</u> been carefully recorded. Wilhelm criticizes Bringhurst (his only mention of Dr. Bringhurst in a footnote) for "erroneously reporting that strawberry breeding began at the Davis station in 1925–26 and has continued there to the present," facts which were reported by Dr. George M. Darrow in his 1966 book, *The Strawberry, History, Breeding and Physiology*. Wilhelm notes that the strawberry breeding program was shifted from Berkeley to Davis in 1945, under the direction of Warren Tufts, following Thomas' departure. Thomas resigned from the university to assume directorship and pathologist of The Strawberry Institute located in Morgan Hill, organized by E.F. Driscoll. There were no other references to Bringhurst or Voth. Wilhelm also notes that, "Foundation stock of *all* University strawberry selections was transferred to the university's Wolfskill Experimental Orchards near Davis." *If this were the case, what planting stock did the Institute utilize for <u>its</u> breeding program?* As mentioned earlier, the Institute breeding program developed varieties which gave Driscoll a competitive advantage until 1952–53, when Bringhurst and Voth bred their first hybrids, the Fresno and Torrey, finally selected in 1961, closely followed by Tioga in 1963. Voth's earlier work in southern California showed that the Lassen variety had tolerance to both mild winters and salinity, but lacked adequate firmness for fresh shipment. These varieties were the first major varietal transformations since the 1945 university varieties and marked the *beginning of the southern California* berry industry and represented 60% of total acreage in 2004. During the 1960s as the summer planting system became dominant with Tioga in southern California, the summer planting system was successfully applied to the Shasta variety in northern

California. Nonetheless, Driscoll varieties continued to dominate the marketing of the northern Shasta because of quality, size, appearance, and summer and fall fruiting. The Driscoll star performer, Z5A (Goldsmith), was an everbearer, not developed for early fruiting, but produced excellent fruit continually from May to November.

The southern California contributions to the development of the day neutral varieties were derived from hybrids developed at South Coast Field Station in 1965. The presently named day neutral varieties are hybrids from Predom, namely 65.63-601 and 65.65-601, or second and third generations of these selections, such as Tristar and Tribute, which were introduced by the USDA, and Fern, Selva, Irvine, Seascape and Capitola, introduced by the University of California. The main difference between day neutrals, developed in California, and everbearers, developed in the northeastern part of the United States, is that day neutrals have a lower temperature threshold of growth and continue to grow in Southern California and produce fruit going into the winter months of October, November, and December and continuing. The everbearers developed in the northeastern states will not vegetate during this period or produce fruit continually until the temperature becomes warmer.

From 1984 to 2001 the day neutral, Selva, successfully replaced the 1979 short day variety, Pajaro, which did not have everbearer fruiting potential, but was superior to Shasta, the university variety. In 2001, another day neutral, Diamante, became the dominant northern California variety. These varieties improved the ability of non-proprietary growers to compete in the market place throughout the May–October period. The lack of color on the interior and exterior of these varieties caused them to be discounted in the fresh market and undesirable for the processors. For this reason and the introduction of new, successful proprietary day neutral varieties in southern California planted in June and producing in the fall months, northern California acreage became stagnant at 12,000 acres and Diamante acreage decreased by 500 acres since its 2003 peak of 6,100 acres. Proprietary varieties increased by 1,200 acres during the same period. The other competitive reason for the Diamante decline and proprietary increase was the new concept of adapting day neutral varieties to southern California, which began with Voth's selection of the Irvine variety. The Irvine had the most adaptation for producing fruit with three planting systems: Summer planting, mother planting (runner set), and Frigo plants planted during July and green plants (plants dug with 4 leaves) planted on August 15th. These plants produce fruit during October, November, December, and continue until inclement weather. This new cultural system caused another revolution in growing and marketing. The Irvine variety had a lower threshold for growth and was the only day neutral everbearer that had adaptation for Oxnard south. Unfortunately, the Irvine had little tolerance for moisture, heat, and did not ship well. Therefore, since university pomologists had been unable to breed an improved, adaptable day neutral variety for Oxnard, progress was left to the proprietary researchers who have developed successful varieties, performing as Voth had conceptualized. As a result, the proprietary acreage in Oxnard in 2005 has increased by 2,000 acres, to 59% of the districts total, and from 15% to 20% of the entire state. This development, and the proprietary success in breeding short day types for both northern and southern California, threatens the relationship between the university, the CSC, and the industry's non-proprietary growers. Since 2001, the proprietary share of the industry has increased from 27% to 37% in 2005, presenting a new set of problems for the CSC and the university (See Tables 2 through 5). As patent royalties and research income continue to decline, growers become less dependant on the CSC and the university partnership for pomology; there could be, even should be, a grower outcry for varietal improvement or reduced assessments. The last chapter on the Supply variable will establish the model and barometers for measuring the CSC research programs as well as suggestions for change.

Fortunately, the development of the early fruiting short day types was the industry growth vehicle until the day neutral development outlined above. The desire for earliness coincides with spring and early summer demand, prior to supplies of deciduous fruits, melons, and local fruits in the United States and other northern hemisphere areas. Bringhurst and Voth understood this, as did proprietary researchers, and thus the quest for varieties with color, size, taste, firmness, and disease resistance from both soil and pest. Realizing the Lassen shortcomings, Bringhurst and Voth developed a long series of new, improved varieties, followed by Shaw's and Larson's Ventana, a partial replacement and improvement over Camarosa, and Albion, an apparent substitute for Diamante, with superior color and taste. The following is a list of new university varieties and release dates:

1. 1961—Fresno, Torrey
2. 1963—Tioga
3. 1972—Tufts
4. 1975—Aiko
5. 1979—Douglas and Pajaro
6. 1983—Chandler, Selva and Parker
7. 1987—Oso Grande

8. 1989—Irvine
9. 1990—Seascape and Capitola
10. 1993—Camarosa
11. 2000—Diamante
12. 2003—Ventana
13. 2004—Albion

Although Tioga originated as a summer planting, the major revolution of annual winter planting began with its high elevation nursery planting in the 1960s. The Tufts variety, a 1963 hybrid from Tioga and introduced in 1972, was found to be the most responsive to nitrogen fertilizers. The preplant fertilizer placement of slow release materials showed the importance of banding as close to the plant as possible in the planting slot. This made Tufts as early and as productive as Tioga, which it finally replaced because of better fruit size.

The Aiko was released in 1975 and was used as a weak day neutral in northern California to provide production from May to October. Its undesirable characteristics of *cat-facing*, light color, poor pollination, and short season offered no competition to the proprietary varieties and highlighted the urgent necessity for an adaptable, competitive day neutral with color, taste and long season characteristics. As an active participant in this variety drama, both as an executive at Naturipe Berry Growers and Chairman of the CSAB, I recognized that our competitive survival and that of all non-proprietary growers was dependant on the breeding of a day neutral variety for northern California. Bringhurst and Voth understood the urgency of the problem but neither has ever complained of the university's failure to prepare for an adequate transition caused by the abrupt departure of Thomas and Goldsmith. Since 1958 and throughout my 50 years in the strawberry industry, I have always believed that The Strawberry Institute possessed planting stock unavailable to the university and additional knowledge on the everbearer, which only they possessed. Meanwhile the quest continued as the Douglas and Pajaro, released in 1979, were selected at South Coast Field Station. The Pajaro, with very little winter planting adaptation, was successfully summer planted in northern California and became the dominant variety because of its exceptional color and fruit quality, yet inadaptable for summer fruiting. It was also a popular freezer variety because of color and taste.

The Douglas variety, especially adopted for winter planting, took over from Santa Maria south during the early 1980s replacing the Tufts variety. It was an exceptionally early variety with a large, conically shaped fruit and deep red color perfect for fresh and freezer use. These two successful varieties, Douglas and

Pajaro, caused production per acre to climb from a low of 5–6 tons in the 1950s, 12–17 tons in the 1960s, 18–23 tons in the 1970s, to a high of 26 tons per acre in 1982. A major reason for this tremendous increase was the 1981 return of the nursery fields to Macdoel, California, where the higher elevations provided earlier chilling, fruiting, better fruit quality, size, and firmness, even with the Lassen variety. Unfortunately, Douglas had a tender skin and did not perform well under heat conditions in the growing areas or markets. Pajaro was not a day neutral and suffered the defect of early fruiting and a short fresh market period compared to the proprietary varieties.

As a result of continual grower and CSAB pressure on Bringhurst and Voth, as well as their professional desire for additional success and improvement, the day neutral Selva and the short day Chandler variety were released in 1983. The Chandler quickly became the dominant variety from Santa Maria south while the Selva finally provided a day neutral alternative to the proprietary varieties. The Chandler had a higher cold requirement than Douglas, but was not as early. The industry required a variety with increased firmness and improved shipping ability for the fresh market as well as being acceptable to the processing industry. To solve the early production defect, the growers followed the old adage, "you have to plant early to be early," which was necessary for the Chandler variety. High elevation plants from Macdoel gave better performance than plants from McArthur and Susanville. Selva appeared to solve the 25-year-old varietal inability to produce from May through October or November, and provide a genuine industry alternative to the Driscoll everbearer in the Salinas-Watsonville area. Due to the success of both the Chandler and Selva varieties, state acreage rose from 13,000 to 20,000 acres by 1990, when the short day Camarosa was introduced. The beloved Chandler, because of an unfortunate early season decrease in size, was quickly relegated to the dust heap of discarded varieties, and the Camarosa, with color, size, taste, and greater production capabilities, caused southern California acreage to increase from 6,500 acres in 1990 to 14,000 in 1993.

A period of varietal stability was experienced from 1993 to 2000 until Drs. Shaw and Larson, the new generation of pomologists, developed the day neutral Diamante and the short day Ventana. Almost overnight, the Diamante took center-stage as a larger, firmer, better tasting and producing than the Selva. By 2003 the Diamante planting rose to 6,100 acres, 52% of northern California acreage and 22% of the state's total 28,230 acres. This variety provided the nonproprietary growers with an acceptable marketing alternative to proprietary varieties and the Selva.

The short day Ventana has increased to 4,800 acres in 2005, from 4% to 15% of the state total, and has proved to be earlier than the Camarosa with better size, shape, and less crooked fruit because of better pollination. The negative aspects of the variety are lack of color, taste, freezer acceptability, and an excessively heavy, almost unmarketable peak and down-sizing, as compared to Camarosa. It is also slow to develop roots at the nursery for early digging and grower planting.

To prevent the same problem experienced by Bringhurst and Voth in 1945 resulting from the abrupt departure of Thomas and Goldsmith, industry leaders and the CSC selected Drs. Doug Shaw and Kirk Larson in the mid-eighties to allow for sufficient training and eventual replacements of Bringhurst and Voth. The CSC, separately from the university, financed the hiring of Dr. Shaw prior to the Bringhurst retirement in order to insure that pomology knowledge and experience was transferred. The university was reluctant to replace Voth after his retirement for financial reasons; CSC pressure on the university facilitated the employment of Dr. Larson, but after some delay.

The quest for the "perfect" berry continued. When Voth was questioned if he could foresee strawberries the size of oranges, he commented, "We would have to change our whole system of packing and processing." In a visionary voice and eyes twinkling in his boyish face, he added, "But, maybe we could do that."

The creation of The Strawberry Institute in 1945 initiated the development of many additional research organizations within the grower/shippers in California, in addition to separate private breeders who sell proprietary varieties to interested individuals and/or grower groups for a per plant fee for investor participants. These varieties can then be marketed by participating organizations. The source of this proprietary planting stock is a problem for the university since they have patent rights over their materials. The proprietary growth, from 29% of total state acreage in 2001 to 38% in 2005 and in Oxnard from 13% to 20.5% during the same period, poses a potential patent revenue decline, which could impact the CSC-university relationship. (See Tables 2 through 5) The same trend toward additional proprietary acreage, in Santa Maria, from 2% to 3% of total state acreage and in Salinas/Watsonville from 12.6% to 14%, augments the potential revenue decrease for the university. The percentage figures mask the startling 2,500-acre state increase in proprietary varieties, a 25% gain from 2004 to 2005 and a 40% upward move in Oxnard. Voth's day neutral research with the Irvine variety provided the opportunity for proprietary growth and success, unleashed a potential for continued stagnation or decline in the northern California industry, a reduction in university revenues, and perhaps undermined the CSC-university relationship and, ultimately, the demise of the independent grower.

The private nurserymen, who supply mostly university materials, will be impacted by this development and forced to reduce their operations or become a contract supplier for proprietary growers and shippers. The CSC should analyze the unintended consequences of this research and plan for varieties that would reverse the trend toward proprietary varieties developed by private, corporate organizations at the expense of the independent grower to whom the CSC and the university are responsible. These growers must pay royalties to the propriety organizations in addition to CSC assessments for research. The CSC board of directors should address this double payment. If this trend continues the assessment rate should be lowered and the reliance on the university diminished.

However, another alternative would be outsourcing pomology to competing research organizations in order to bring competition into the breeding program, a suggestion that was made 10 years ago. Although some progress has been made with the Ventana and Diamante, much more must be done because Ventana has a short marketing period, Diamante has white flesh, and there is no competitive variety for Frigo planting in Oxnard. The CSC-university model is not the only model for research; the CSC and all commodity boards should consider alternative models.

Table 2
Acreage Survey Results by District: 2001–2005

By District	2001	2002	2003	2004	2005	Change	% Change
Orange Co. &							
San Diego	2,446	2,538	2,883	2,899	2,457	(442)	-15.2%
% State	9.7%	9.5%	10.2%	9.2%	7.5%		-17.8%
Oxnard	7,777	8,582	8,794	10,349	11,333	985	9.5%
% State	30.9%	32.0%	31.1%	32.7%	34.7%		6.2%
Santa Maria	3,817	4,100	4,438	5,647	6,293	646	11.4%
% State	15.2%	15.3%	15.7%	17.8%	19.3%		8.0%
Watsonville	10,759	11,300	11,687	12,201	12,250	49	0.4%
% State	42.8%	42.1%	41.4%	38.6%	37.5%		-2.7%
San Joaquin	344	309	429	544	304	(240)	-44.1%
% State	1.4%	1.2%	1.5%	1.7%	0.9%		-45.8%
State Total	25,143	26,829	28,230	31,639	32,636	998	3.2%

Table 3
2005 Acreage Survey Planting Dates by District

By District	Summer	Winter	Total
Orange Co. & San Diego	30	2,427	2,457
% State	0.7%	8.6%	7.5%
Oxnard	3,300	8,033	11,333
% State	76.5%	28.4%	34.7%
Santa Maria	429	5,864	6,293
% State	9.9%	20.7%	19.3%
Watsonville	253	11,997	12,250
% State	5.9%	42.4%	37.5%
San Joaquin	304	0	304
% State	7.0%		0.9%
State Total	4,316	28,320	32,636

Table 4
Varietal Acreage Trends: 2001–2005

By Variety	2001	2002	2003	2004	2005	Change	% Change
Camarosa	10,931	10,485	9,626	9,832	6,462	(3,370)	-34.3%
% State	43.5%	39.1%	34.1%	31.1%	19.8%		-36.3%
Diamante	5,141	6,408	7,445	7,527	7,515	(11)	-0.1%
% State	20.4%	23.9%	26.4%	23.8%	23.0%		-3.2%
Other	249	455	533	659	1,610	951	144.2%
% State	1.0%	1.7%	1.9%	2.1%	4.9%		136.7%
Proprietary	7,385	8,068	8,421	9,756	12,271	2,515	25.8%
% State	29.4%	30.1%	29.8%	30.8%	37.6%		21.9%
Ventana	0	62	1,020	2,777	4,779	2,002	72.1%
% State	0.0%	0.2%	3.6%	8.8%	14.6%		66.8%
Total Acreage	25,143	26,829	28,230	31,639	32,636	998	3.2%

Table 5
Organic Acreage by District: 2001–2005

Organic by District	2001	2002	2003	2004	2005	Change	% Change
Orange Co./San Diego	48	52	70	76	48.1%	(28)	-36.8%
% State	0.2%	0.2%	0.2%	0.2%			-38.8%
Oxnard	62	60	232	153	93	(60)	-38.9%
% State	0.2%	0.2%	0.8%	0.5%	0.3%		-40.8%
Santa Maria	14	0	0	0	0	0	0.0%
% State	0.1%	0.0%	0.0%	0.0%	0.0%		0.0%
Watsonville	181	271	306	378	399	22	5.7%
% State	0.7%	1.0%	1.1%	1.2%	1.2%		2.5%
State Total	305	383	607	607	541	(66)	-10.9%

Horticulture

Wilhelm chronicles the research effort of the University of California begun in 1924 by Anthony G. Plakidas on the disease xanthosis, found in the Banner variety. The new disorder was first called strawberry blight and it spread through entire fields within one year affecting plants and runners. Growers had previously avoided disease by moving annually to new ground. Research indicated that disease free plants resulted in clean stock and pointed to healthy stock as a factor in control of disease. Research by plant pathologists at Berkeley began a breeding program as a means of disease control. Prior to the 1945 development of the 5 university varieties, this research for clean stock led to the discovery of Verticillium wilt, which seemed to be soil borne. This began the research into pests and other soil diseases, which caused parent strawberry lines from being healthy and full fruiting. The primary goal was disease-resistant strawberries, which resulted in the hybridization of thousands of crosses and climaxed in the Thomas-Goldsmith 5 university varieties.

In addition to breeding for disease free lines, a Strawberry Plant Certification Program was developed by the State of California Department of Agriculture Nursery Service in 1940 to enforce the responsibility of nurserymen to provide healthy, true to type plants and legal status to strawberry virus, nematode, and other diseases. The program marked full recognition by the industry of these important guidelines and provided for technical inspections to accomplish its objectives. In the 1970s, a meristemming program to insure virus free nursery plants was developed on a small, part-time scale at Berkeley with Ruth Mullin in charge. The program was poorly funded until the CSC insisted that the university increase and improve its clean stock program. The Foundation Plant Material Service undertook the clean stock research program, which was then transferred from Berkeley to Davis (See last chapter on Supply Variable).

Verticillium wilt was becoming a significant problem, as were other soil-borne diseases. The Strawberry Institute, Wilhelm, and Voth at the South Coast Field Station were now considering the elevation levels for nursery stock and chilling requirements. These problems developed prior to the industry creation of the CSC, the hiring of Victor Voth in 1946, and Bringhurst in 1953. The agenda for the university was clear as the primary researchers, Thomas and Goldsmith, departed. Voth and Southern California growers realized that the university variety Lassen had tolerance to both mild winters and salinity but lacked firmness. The first varieties with additional firmness were developed in 1952–53 and the next varieties Fresno and Torrey were selected and named in 1961. The long list of hybrids that followed was listed earlier.

The horticulture revolution was set to begin. Prior research by Plakidas, Thomas, Goldsmith, Wilhelm, and other University of California researchers provided the environment and background for the major, innovative work accomplished by Voth and Bringhurst, soon to be funded and supported by the industry creation and funding of the CSC.

Strawberries are considered to be one of the most sensitive plants; therefore it was necessary to minimize the effect of salinity. Furrow irrigation, the method of delivering water to the plant, exacerbated the salinity problem and reduced fruiting and the health of the plant. The first experimental work began to show the use of sprinkler irrigation in all southern California and the positive effect on reducing salinity. This irrigating system was also the beginning of pre-plant land preparation, irrigation after planting, as well as the beginning of the development of summer and winter annual planting systems, soil fumigation, drip irrigation and utilization of polyethylene mulch for weed control, plant stimulus, and cleanliness of fruit.

In 1957–58 experimental work with polyethylene mulch began. At that time the eastern seaboard was experimenting with black mulch that found some use in southern California. At the South Coast Field Station work began comparing black and clear mulch, showing that *clear* mulch increased soil temperature at least 10° Fahrenheit during the short days, thus enhancing the production of early fruit and more total fruit, especially when annual planting was used. Simultaneously, Dr. Stephen Wilhelm worked on Verticillium wilt and found that soil fumigation with chloropicrin gave *control plus plant response* for the *strawberry replant problem.* When clear polyethylene mulch was used, weed control became mandatory. Experimental work began with methyl bromide under mulch for weed control, in combination with chloropicrin. Growers adopted this practice because the methyl bromide gave adequate weed control for use with clear polyethylene mulch and showed effective early results. Chloropicrin controlled most disease problems with a plant response of about 20%–40% increase in yield and meanwhile overcame the replant problem caused by soil-borne disease. The combination of clear mulching and soil fumigation was a fortunate combination for this revolutionary system, in conjunction with sprinkler irrigation, and set the stage for experimental work in drip irrigation technology. In 1967 drip irrigation experiments began using a system of emitters every two feet that showed less water was used, especially when mulch was applied, due to less surface evaporation and with less salt accumulation. Dick Chapin, Chapin Watermatics, developed the first drip tape in 1970 using the twin wall method, which proved that *good water placement* facilitated all field operations. Simultaneously, *new bed shape* experiments increasing plant population were conducted at the South Coast Field Station showing another advantage of drip irrigation, which gives each plant more space with a *four row bed with two drip lines.* This system increased yield on winter planting, with the same plant population, 30%–50 %. Voth's innovation was adopted in 1972 for the first time in Oxnard and is currently used for winter plantings in all areas south of Santa Maria.

Another integral part of this planting system was the experimental preplant use of slow release fertilizers, banding as close to the plant as possible in the planting slot. The Tufts variety introduced in 1972 was found to be the most responsive to the nitrogen fertilizer slot. The new irrigation method and fertilizer placement, described above, proved that water placement determined fertilizer placement, and *fertilizer placement was more important than the material used.*

These major horticultural innovations revolutionized the strawberry industry and are in practice today. However, the story of these two "iconoclasts" is not yet concluded. Thomas and Goldsmith began important work on the effect of nursery elevations, preplant chilling requirements, and the conceptual development of optimum planting dates before their university and industry work ended in 1944–45, when both left to form The Strawberry Institute with Driscoll. The cultural practices at that time, and until the above innovations, included the harvest of fruit from plants 2 to 3 years old. The development of annual planting, suggested by Voth in 1955, was adopted after the disastrous strawberry year in 1957 after which 20,700 acres in California (70% processing, 30% fresh) dwindled to less than 9,000 acres by 1962. Annual planting did more to minimize insect and disease problems of the strawberry than any one single pesticide intro-duced since 1955. A host free period in any given area gives the best control for any pest and disease; Voth proved this principle in southern California when very little or no virus problem had been encountered with the use of new, clean plants each year. Thomas and Goldsmith appreciated the need for disease free nurseries and plant breeding grounds but did not foresee annual planting. Their planting system included spring planting and fruiting by July of the same year from blos-soms carried by the plant when it was set in the ground. Production was relatively small and the first major crop was not harvested until the following spring. The plants carried over during the winter encouraged disease and pest development, because there was no host free period. To make matters worse, these same plants were kept one or two extra years, exacerbating the infestations and leading Voth to annual planting of early fruiting varieties to be disked under in July, allowing the same land to be prepared for replanting in September and October. While the practice of early, or winter, planting between November 15 and February 15 was recommended by J. B. Cutter back in 1916, it was not adopted commercially until 1941 by Ned Driscoll. This concept of early and annual planting, which was the beginning of the revolution to follow, resulted in some fruiting during the first year in the ground and the major crop producing during the second year with over wintering. The system began the practice of "planting fields solid," thus saving a year of tedious *catching down runners* from mother plants (planting runners between mother plants) set several feet apart on the beds. This new practice caused a considerable change in the nursery system of supplying plants for commercial growing. In order to supply the increased demand resulting from solid setting and to provide plants with additional nursery chilling, nursery acreage had to be expanded and moved to higher elevations, from the Sacramento Valley to the northern higher desert areas. The additional chilling was required for the earlier dormancy necessary for earlier nursery digging and for earlier commercial planting and fruiting production. The cold, which was required for dormancy, was partially aided by low temperature storage.

Bringhurst and Voth realized the importance of earlier chilling and dormancy than had previously been considered. Since 1955, research on annual winter planting had continued in southern California and by 1958–59 planting of high elevation-high latitude nurseries began, thus providing for earlier planting in the commercial growing areas. Voth and Bringhurst stressed the advantage of straw-berry plants produced at higher latitude and elevation for early fruit production, especially in southern California. This was clearly shown in early data, 1957–1958, from Torrey Pines. The plants produced in areas with the most con-sistent accumulated hours of temperature below 45° F have the most stored energy in the crowns and roots, especially during August, September, and October. This gives the plants more energy to withstand the rigors of digging, handling and planting in addition to better growth at lower temperatures, thereby allowing the plants to be dug and planted earlier. Macdoel plants were shown to be superior to all others due to higher latitude and elevation, but profitable numbers of plants per acre were insufficient. During the early 1970s, the Macdoel area was discontinued while Fall River and McArthur became the dominant high elevation nursery locations. With the introduction of the Tufts in 1972 and Douglas in1979, the annual winter planting system became more dominant from Santa Maria south. Nursery operations were moved from Fall River-McArthur to Susanville because of the lower chilling requirement of the Douglas variety, which provided more adaptation for early fruiting; it became the largest nursery area for winter planted production.

During 1981 the nursery area of Macdoel resumed, as Lassen Canyon Nursery shifted from lower elevations, as a result of Voth's previous experimental work in southern California in the late 1950s and early 1960s, which proved that this area gave earliness, better fruit quality, size and firmness even with the Lassen variety. Additionally, Lassen Canyon Nursery moved to Macdoel because of lighter soils, which contributed to more efficient harvesting after rains.

Illustrations on the following pages: Harvesting Lassen Canyon Nursery strawberry plants at high elevation nursery at McDoel, CA.

Picture 10. Top: Kenny Elwood, Liz Elwood Ponce. Center: Kenny, Jr and Kenny, Sr. Bottom: Roger Hamamura, Herb Baum, Victor Voth, Kenny Elwood.

*Picture 11. Top: Cutting off plant tops. Center: Kenny Elwood.
Bottom: Bagging cut plants for packing shed.*

Picture 12. Top: Cutting machine. Bottom left: Digging machine.
Right, Roger Hamamura inspecting plants before packing.

The introduction of Chandler in 1983, with a higher cold requirement than Douglas, was not early and caused the growers to follow the adage, "you have to plant early to be early," which was necessary for the Chandler variety. Plants from Macdoel gave better performance than plants from McArthur and Susanville.

Bringhurst and Voth emphasized the important relationship between temperature and respiration rate for winter planting and Frigo plants (summer planting), both at the nursery and fruit producing areas. Plants should be cooled down as soon as possible after digging because dehydration and temperature are most important in plant handling. Winter plants should be cooled to +1° C or 34° F and Frigo plants, to a minimum of 28° F or–2° C to keep the respiration rate to a minimum.

Important environmental factors for conditioning plants for optimal fruit production at nursery locations would be a latitude-elevation harvest date and plant chilling for winter planting at 34° to 36° F or 2 C, and for fruit production, planting date latitude, altitude and temperatures for optimum quality and continuous production should be a daytime maximum of 75° F and nighttime minimum of 55° F. From these physiological principles, Voth makes certain statements:

1. The more production early, the more production late as well as total production. If planting is too early, fruit size, quality and firmness will be sacrificed.

2. All strawberry growing areas benefit from this principle: The annual planting system suggested in 1955, when summer planting was introduced proved the adage, "The younger the plant the fruit is produced on, the better the size, flavor, appearance, and firmness."

3. The success of short day varieties depends on how active the plant grows during November, December, and January, which sets the pattern for each production year.

4. All everbearer and day neutral plants are highly sexual and benefit from chilling more than short day types; the sexual exploits the asexual (vegetative) reducing size and quality. The Selva variety gained the most from chilling, as observed in the Santa Maria area, when Frigo plants planted in April gave good quality during late summer and fall. Selva also gave better quality early in the Salinas-Watsonville area when plants received optimum chilling in the nursery and had correct planting time for any given location.

This study would be incomplete if we did not include the material provided by Victor Voth on the important subject of pest and disease control:

1. The Entomologist and Plant Pathologist spend too much time on the pest and disease and not enough time on integrating a control program for a given area.

2. In California during the winter months, optimum pest and disease control is mandatory for optimum growth during the winter months in all areas. Pests and disease must be controlled early and prevention must depend upon cultural methods and a spray program. It should be called integrated *crop* management not integrated *pest* management.

3. All during and after planting, the plant is differentiating leaves, fruits, buds, blooms, and spray programs must consider all of these processes. Spray concentrations are very important because blooms are most sensitive; high gallonage, low concentration sprays are preferable and low volume, high concentration sprays should be avoided. Preventative measures should be taken early to avoid high infestations of pests and disease, which then may require materials with *phytotoxicity* effecting bloom and fruit.

The Voth-Bringhurst revolution was overwhelming. The descriptive model in Chapter 5 evaluates the influence on the variables which affects the supply of strawberries, varieties, costs, yields, and horticultural innovations. Also, the effects of horticulture and pomology on supply and demand for strawberries are analyzed in the model, as the changes influence the variables included in that model.

The CSC gradually increased its mandatory assessment rate from ½ ¢ to 5 ¢ for each fresh and freezer crate during 1970–2003 to help finance this revolution in pomology and horticulture. When Bringhurst retired in 1990, the university assumed the Shaw expense.

Marketing

As indicated in Chapter 2, an important purpose for the formation of the strawberry commodity board was to enhance the competitiveness of the California strawberry industry in the national and international marketplace. As previously indicated, the mandatory assessment rate was increased to implement programs to solve the marketing challenge created by the doubling of California acreage and growth in per-acre yields. The CSC responded to the volume and price challenge, as well as the problem of inadequate varietal quality, with major expenditures on pomology, horticulture, and marketing. The objective of the CSC was to increase demand by directly influencing the consumer, similar to the media efforts of major food and non-food firms. This also required the simultaneous non-proprietary varietal improvement in strawberry quality, taste and shelf life, while extending the season. The advantage gained by the Driscoll-Thomas-Goldsmith proprietary breeding program, with berries of superior quality, taste, shelf life and extended season, provided the emergency environment required for *urgent* industry development of new marketing and research expenditures through the CSC.

In addition to research in the above areas, simultaneous research on the problems of transportation, precooling and packaging were undertaken, again with the CSC in partnership with the University of California, industry growers and shippers. Before World War II and until 1966, rail transportation was the mode for all fresh produce shipments east of the Mississippi River. All berries were shipped in railway express cars on separate trains or with mail and other express freight on expedited passenger rather than freight schedules. Refrigeration was provided by wet ice and dry ice with salt added at point of origin and in transit, as per prior shipper instructions. Another revolution occurred in 1966 as a result of the post-war interstate highway system, which facilitated rapid, affordable truck delivery of fresh berries anywhere in the country using mechanical refrigeration. The reduction in transportation time to markets, superior refrigeration, and convenience of direct delivery, resulted in improved quality and longer berry shelf life. Gordon Mitchell of the University of California provided the research, with grower/shippers cooperation in the 1960s and 70s, for the major breakthrough in rapid pre-cooling. Prior to this development, berries were stored and cooled overnight, then loaded the next day. In many cases, cooling was inadequate causing quality deterioration. Rapid pre-cooling to a low of 34°, slowed down the respiration rate, and permitted same day shipment providing longer shelf life and improved quality. Heggblade and Marguelas, once a major California grower/shipper, began this revolutionary pre-cooling procedure in the 1950s in Anaheim, with rapid pre-cooling in railway express cars using custom designed refrigeration equipment. This equipment was also used later in the harvest season in Arroyo Grande north of Santa Maria.

Palletization became part of the new rapid cooling system, as hand stacking in the field and restacking again at the cold storage plant was replaced by field palletization. Field palletization enabled the application of a process using a product called Tectrol©, in which CO_2 gas is injected into a pallet of berries enclosed with shrink-wrapped plastic. Research at shipping point and destination retailers proved longer shelf life and less decay was the result of the Tectrol© process.

Cardboard cartons began to replace wooden crates after WWII and by 1960 wooden baskets with metal tops were replaced with plastic pints, both of which provided added air circulation, improved quality and provided retailers with more opportunity for merchandising and advertising. Most berries are currently packed in 1-lb, 2-lb, or 4-lb plastic containers with barcodes for easy scanning and product protection. This change allows better retail inventory and reorder control. Like with all revolutionary changes, growers and shippers resisted, but the transition was successfully accomplished with the help of university support, Gordon Mitchell's research, and industry field and shipping trials.

Illustration on the following page:

Picture 13. The Paine and Shimizu families pioneered the wooden and plastic basket container industry.
Top: Tom Paine, Alan Shimizu. Middle: 1955 wooden strawberry basket. Bottom: Some of the current packing options, such as full trays, half trays, 1 and 2 pounders.

The CSC marketing activity mirrors the revolutionary innovations in pomology and horticulture. *Three distinct CSC phases of economic activity* mirror the revolution, which took place in pomology and horticulture: Phase 1, early years; Phase 2, media years; and Phase 3, mature years. Phase 1, early years, began with no advertising, little promotion, and no field representatives to make marketing presentations to retailer or food service operators. In 1957, with production at a low of 5.4 tons per acre on a total of 20,000 acres, the long decline in acreage began (See Table 6). With the development of plastic mulch, soil fumigation, and drip irrigation, yields began rising, and acreage, which hit a low of 7,800 acres in 1966, began to climb. Yields had tripled by 1970 and the new Tufts hybrid improved quality and profitability. Grower profitability and per-acre yields increased and acreage expanded from 7,800 to 12,500 acres during the last half of the 1970s. These increases caused California production to double between 1970–1980, creating a real marketing challenge for the industry and the CSC. This was the beginning of Phase 2 of economic activity. To respond to this challenge, a major advertising program began in 1974 with the objective of increasing demand. The goal of the CSC was to influence the consumer, similar to the major food and non-food firm's usage of the mass media to accomplish this objective. CSC attempted to model a program of advertising and merchandising similar to Sunkist Growers, the only major fresh produce firm involved in this activity. During the period of 1974 to 1976, when the CSC used a low level of TV advertising to augment demand, the number of markets receiving TV coverage increased from 20 to 58, and this level of advertising appeared to be effective in generating retail merchandising but not consumer purchases. With continually rising crop production, it appeared necessary to move into a new intensity of conscious consumer advertising. In 1976 the CSC tested heavier weighted advertising in 2 controlled markets and, based on the research results, the Board voted to begin TV advertising in 14 key markets and to gradually increase the number of markets. Low-level TV advertising continued in 44 markets in conjunction with merchandising efforts, even though research indicated little difference in market unloads between controlled markets. This led to irrational concentration on heavy weighted TV until 1989 when further research indicated the expense of direct consumer TV continuation was prohibitive. The evaluation of these programs using the barometers in the descriptive model is found in Chapter 6.

The media years of Phase 2, 1976–1990, was clearly the most dynamic period of economic activity of the CSC, responding to 1,000–acre annual increases. TV, radio, tie-in ads, display contests, and public relations were the programs used to influence the consumer, increase demand, price, and grower returns. These items increased budgets by 50% while horticulture and pomology expanded from 7% to 13% of the budget. This trend continued until 1989, when consumer advertising and merchandising peaked at $1 million and $639,000 respectively, or a combined 70% of the budget. By 1990, CSC had realized the ineffectiveness of the media expenditure and spent only $300,000. Meanwhile, promotion activity designed to encourage retailer advertising had increased to $1.2 million becoming the leading strategic plan for increasing demand, based on the assumption that retail advertising does, in fact, increase consumption. Media studies will be evaluated in the discussion of the variables as the effects of alternative programs are measured.

During the 1980s mergers and expansions caused retailing to become more concentrated with only 102 supermarket chains reporting annual sales of over $100 million. Fewer but larger supermarkets began to sell 77% of the industry's $179 billion total volume in 1980, a trend that has continued.

The expansion of the food service industry also accelerated during 1970–1980 to $112 billion and rivaled the supermarket industry as a major distributor, while consumers, who spent 40% of the food dollar, turned increasingly to out-of-home dining. In 1979, 25% of the strawberry crop was distributed to the food service industry.

In the 1980s, acreage continued to increase from 11,000 acres in 1980 to 20,000 acres in 1990. Yields per acre rose to 27.6 tons in 1990 from 20 tons in 1979 and production soared to 1 billion tons from 500 million in 1980. Meanwhile per capita consumption doubled from 1.7 lbs in1970 to 3.6 lbs in 1991. This active, second phase of economic development greatly expanded CSC income_requirements and the grower mandatory assessment rate rose to its maximum 5¢ per tray, as grower/shipper/processors sensed a crisis of excessive production, and as anticipated, peak volume reached 2.5 million 10 lb trays per week, with a record 500,000 trays shipped on various days. By 2002, the season peaked in excess of 5,000,000 trays per week, with a series of 1 million tray days.

Table 6
Processing Strawberry Advisory Board of California – California Strawberry Production, Acreage Yield, Farm Value: 1950-1976

*000 Omitted
**Includes Juice Berries
***Fresh Value based on Weighted index of FOB prices quoted by Federal-State Market News (Not actual prices after adjustments, consignments and other factors affecting actual price) Freezer Value – FOB Processing plant

Year	Acres	Total Production Pounds*	Yield Per Ac. Tons	Marketed Fresh Pounds*	Marketed Freezer** Pounds*	Mktd. Fresh % Total	Mktd. Freezer** % Total	Farm Value*** Fresh	Farm Value*** Freezer	Farm Value*** Total
1950	5,700	81,282	7.1	47,724	33,558	58.7%	41.3%	9,545	6,712	16,257
1951	6,900	88,113	6.4	54,317	33,796	61.6%	38.4%	11,570	6,759	18,329
1952	8,400	114,912	6.8	68,162	46,750	59.3%	40.7%	13,632	7,480	21,112
1953	9,400	152,938	8.1	71,712	81,226	46.9%	53.1%	13,840	13,402	27,242
1954	10,900	159,467	7.3	71,577	87,890	44.6%	55.4%	16,176	14,062	30,238
1955	14,000	166,740	6.0	64,400	102,340	38.5%	61.5%	16,229	17,398	33,627
1956	19,000	243,200	6.4	88,500	154,700	36.2%	63.8%	19,558	21,685	41,216
1957	20,700	223,560	5.4	118,260	105,300	52.9%	47.1%	21,169	10,951	32,120
1958	17,000	214,200	5.3	101,200	113,000	47.2%	52.8%	20,341	13,221	33,562
1959	13,200	170,280	6.4	100,208	70,072	58.8%	41.2%	22,144	10,360	32,504
1960	11,700	156,780	6.7	89,233	67,547	56.9%	43.1%	20,673	10,863	31,536
1961	11,500	202,152	8.8	132,503	69,649	65.6%	34.4%	28,408	8,030	36,438
1962	10,500	211,168	10.0	137,760	73,408	65.2%	34.8%	30,124	8,740	38,864
1963	9,800	238,140	12.1	156,284	81,856	65.7%	34.3%	35,275	9,672	44,947
1964	9,000	228,600	12.7	141,124	87,476	61.8%	38.2%	34,428	12,425	46,853
1965	8,300	174,832	10.5	105,598	69,233	59.0%	41.0%	27,950	11,999	39,949
1966	7,800	177,840	11.4	118,145	59,604	66.5%	33.5%	30,743	9,982	40,725
1967	8,000	208,800	13.1	148,034	60,766	70.9%	29.1%	34,359	8,923	43,282
1968	8,600	297,345	17.3	220,453	76,892	74.1%	25.9%	49,247	11,494	60,741
1969	8,400	262,503	15.6	195,800	66,503	74.6%	25.4%	50,600	10,093	60,693
1970	8,500	288,191	17.0	214,661	73,530	74.5%	25.5%	51,050	11,261	62,311
1971	8,300	300,131	18.1	232,983	67,148	77.6%	22.4%	60,865	9,520	70,385
1972	7,800	282,978	18.1	224,716	58,262	79.4%	20.6%	58,411	9,153	67,564
1973	8,100	325,892	20.1	232,604	93,288	71.4%	28.6%	66,196	17,634	83,830
1974	8,900	382,258	21.5	277,307	105,068	72.5%	27.5%	85,223	19,233	104,456
1975	10,000	378,996	18.9	269,859	109,137	71.2%	28.8%	91,835	20,402	113,237
1976	10,800	422,731	19.5	282,614	140,117	66.8%	33.2%	103,726	32,083	135,809

Table 7
Processing Strawberry Advisory Board of California – California Strawberry Production, Acreage Yield, Farm Value: 1977-2004

* 000 Omitted
**Includes Juice Berries
***Fresh Value based on Weighted index of FOB prices quoted by Federal-State Market News (Not actual prices after adjustments, consignments and other factors affecting actual price) Freezer Value – FOB Processing plant
(1) Estimated

Year	Acres	Total Production Pounds*	Yield Per Ac. Tons	Marketed Fresh Pounds*	Marketed Freezer** Pounds*	Mktd. Fresh % Total	Mktd. Freezer** % Total	Farm Value*** Fresh	Farm Value*** Freezer	Farm Value*** Total
1977	11,600	520,888	22.4	342,500	178,388	65.7%	34.3%	$131,255	$ 37,107	$ 168,362
1978	12,900	501,152	19.4	371,297	129,855	74.1%	25.9%	128,185	20,654	148,839
1979	11,500	460,913	20.0	315,860	145,053	68.5%	31.5%	131,293	38,399	169,692
1980	11,000	512,511	23.3	351,886	160,625	68.7%	31.3%	164,874	38,930	203,812
1981	10,900	523,149	24.0	380,685	142,464	72.8%	27.2%	186,617	36,940	223,557
1982	11,600	614,164	26.5	391,906	222,258	63.8%	36.2%	225,574	68,845	294,419
1983	13,000	609,342	23.5	395,708	213,634	64.9%	35.1%	207,933	61,752	269,685
1984	14,100	726,367	25.8	563,686	162,681	77.6%	22.4%	276,372	30,450	306,822
1985	14,600	762,717	26.1	570,320	192,397	74.8%	25.2%	290,428	35,864	326,292
1986	15,600	778,489	24.9	571,167	207,322	73.3%	26.7%	338,068	44,776	382,844
1987	16,800	828,283	24.6	566,615	261,688	68.4%	31.6%	325,186	68,321	393,507
1988	17,650	860,109	24.4	656,377	203,732	76.3%	23.7%	343,082	42,906	385,988
1989	19,900	857,928	21.6	649,215	208,713	75.7%	24.3%	324,518	45,967	370,485
1990	20,000	982,904	24.5	666,519	316,385	67.8%	32.2%	345,649	85,450	431,099
1991	21,100	1,105,141	26.2	773,217	331,924	70.0%	30.0%	384,138	81,377	465,515
1992	24,000	1,026,702	21.2	759,427	267,275	74.0%	26.0%	453,688	59,340	513,028
1993	25,000	1,128,485	22.5	761,982	366,503	68.2%	31.8%	451,445	92,531	543,976
1994	23,300	1,316,888	28.1	890,727	426,161	67.6%	32.4%	523,632	110,494	634,126
1995	23,600	1,208,020	25.5	815,751	392,269	67.5%	32.5%	518,207	91,064	609,271
1996	25,200	1,291,138	25.2	928,270	362,869	71.9%	28.1%	524,423	67,179	591,602
1997	22,500	1,255,619	27.9	895,837	359,782	71.3%	28.7%	593,550	92,868	686,418
1998	24,200	1,304,751	27.0	860,711	458,645	65.3%	34.8%	619,883	136,965	756,848
1999	24,600	1,447,271	29.4	972,437	505,902	67.2%	32.8%	727,610	148,623	876,233
2000	26,300	1,452,046	27.6	1,045,118	421,409	72.0%	28.0%	673,995	93,311	767,306
2001	25,100	1,267,241	25.2	942,191	338,960	74.3%	25.7%	729,731	96,747	826,478
2002	27,200	1,473,645	27.3	1,070,796	431,787	71.3%	28.8%	852,843	137,740	990,583
2003	28,200	1,622,681	28.7	1,182,697	471,455	72.9%	28.5%	925,368	126,124	1,119,116
2004	31,600	1,670,463	26.4	1,224,587	476,391	72.0%	28.0%	1,050,329	118,711	(1)1,169,040

Phase 2, commencing in 1976, included a significant transformation from the *spot market* to what I call a *sales planning* market, or *upside price management*, as major grower/shippers now refer to it. This sales planning concept required each shipper to analyze historical industry demand, volume, and price data, in conjunction with the CSC, then to project proprietary volumes and prices as well as industry volumes directly to retailers. This concept of projected prices and volumes spurred greater sales participation among retailers and suppliers. However, for sales planning to be effective, shippers must provide berries at pre-committed prices. During a 10 year period, 1980–1990, a minimum 10–14 day time lag occurred between announcement and the commencement of the promotion. Shippers frequently failed to project volume accurately or were fearful of committing at lower prices than the spot market dictated at that time. This period was also the era of the small, independent grower, who was part of an independent shipper group or cooperative. These competitive firms needed to return comparable prices to member growers or would be in jeopardy of losing growers. Therefore, the fear of advance price commitments to retailers carried considerable risk and was a deterrent to advance pricing. The industry understood, however, that the failure to price commit and project future supply increases, always set in motion a considerable downward spiral in daily prices, which were inevitably below the current market price necessary to clear the market with the existing volume. The resultant downward spiral usually continued for many days or weeks until a new, higher, realistic equilibrium price was established, based on current supplies. The 10–14 day time lag mentioned above meant then, and today, that until retailers plan and implement ads, which we know will increase demand, an increase in supply will cause a lower level of prices than supplies justify. This time lag requires advance pricing based on anticipated supplies, not current supplies. The concept of sales planning or pre-pricing gradually took hold in the industry and appears to be the main reason for relative price stability in the 1980s, when normative prices averaged 50¢ per lb in 7 of 10 years (See Table 7). The effects of TV and radio are analyzed in the discussion of the advertising and promotion variable, and its effect on demand and price.

Illustrations on the following pages: National television advertising and promotion programs.

Picture 14. Top: Dave Riggs, Bob Reinecke, Bob Kavet, Dinah Shore.
Center: Willard Scott.
Bottom: Gizelle Cahnbley, 1966 Strawberry Queen, Art Linkletter, Willard Scott.

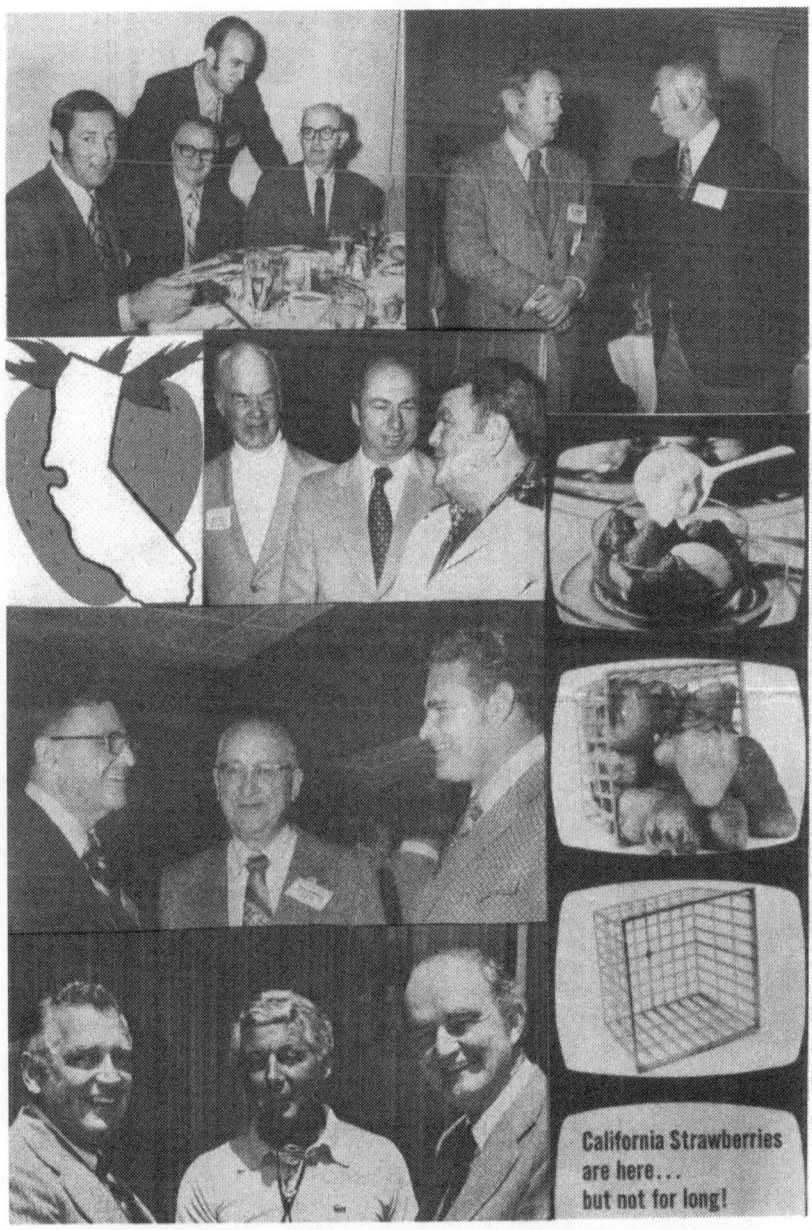

Picture 15. Top: Clint Miller, Peter Aiello, Barney Radovich, Jack Angell, Bill Deardorff, Warren Hinz.
2nd Row: Original California Strawberry Advisory Board logo adapted in 1955,
Harry Russell, Doug Gamble, Bill Crowley. 3rd Row: Herb Baum, Merv Arronchild, Tony Ruso.
Bottom: Bob Kavet, Bill Deardorff, Barney McClure. Right side: TV spot ad from the mid 70s.

Picture 16. Top: Mike Hollister, CSAB Merchandising Director with Field Merchandisers, 1981.
Center: Mike Hollister at retail promotional display for CSAB contest.
Bottom: Dave Riggs and Field Merchandisers, 1978.

The emphasis on sales planning by shippers was vital to the marketing and grower success during the 1970s–80s in conjunction with the CSC's consumer and retail merchandising efforts utilizing TV, radio, billboards, print media, point of sale material, display contests, and a strong public relations program directed toward the food service industry. CSC programs represented new and fresh approaches by the strawberry industry. Per acre yield increases and a tremendous improvement in quality, resulting from Bringhurst-Voth's successful hybrids and innovative cultural practices, played an important part in increasing the demand and availability of strawberries.

Strawberries continued to gain market share in produce departments, increasing from 2.84% in 1981 to 3.7% in1983. Retail advertising activity reversed a 5-year decline in 1983 and increased 33%. The number of ads in April alone was up 226%, down 21.6% in May, and increased 48% in June over 1982. These ad fluctuations, caused by supply availability and suppliers willingness to offer prices and product for promotion, make it difficult to relate CSC or any commodity board to retail ad activity. Many variables are involved in retail ad planning and execution, the most important being supplies, quality, price, and alternative commodities. CSC research showed that customers purchased 76% more fruit in response to a strawberry ad feature than the normal impulse purchase. In 1986, 36% of all consumers were inclined to switch to stores with an ad and retailers sold four times their normal volume with a special feature, which could explain why actual scanner sales data for a major chain indicated a 60% increase in response to an ad compared to the prior week with no advertising.

The food service use of fresh and frozen strawberries continued expanding in the 1980s as the restaurant usage indicated an 89% increase and 75% of all hotels served fresh strawberries, resulting in a usage increase of 52%. Frozen berry usage jumped 90% in the restaurant segment as beverages became the largest user of frozen berries. The food service industry had increased from 20% in 1954 to 48% in 1988, while the retail share had declined from 70% to 56%. A 1989 survey of 200 restaurant owners found strawberries to be the most profitable menu item and 75% of them mentioned strawberries as the top margin builders.

The media years of the 1980s resulted in a period of growth, experimentation, and study. The media was a new tool; strawberries were becoming a major item in the retail stores well as in the food service business. Educating everyone about the growth in the industry, greater supply availability, improved quality and shelf life, was the *goal* of the CSC and industry. The *wearout* concept discussed in the economic literature had not yet become relevant.

The 1990s began a mature Phase 3, ending the trial and error period of the 1980s which had provided an understanding of what the media could accomplish.

Studies with unconvincing data showed that only high levels of TV with 150 or more *gross rating points* (GRP) could affect demand, but the costs were prohibitive. Sales planning, precommitted pricing, and the relatively new *contract pricing* before production became vital tools in the private sector strategy. Rather than trying to influence the consumer directly as in Phase 2, the CSC strategy was to directly establish incentive ad programs with retailers, cooperating with suppliers for ad timing and volume. Shippers provided advance supply and price information. The CSC advertising incentive program began in 1990. Each participating chain was required to submit confidential data to the CSC, reflecting their previous levels of advertising, and incentives were based on prior ads, ad averages for chains nationally, and in each region.

Consumer magazines, public relations with food editors, infomercials, recipe releases, billboards, and radio were utilized in experimental ways during the final marketing phase from 1990 to 2001. Strawberries were making a substantial contribution in the competitive retail struggle for market share and improved margins. The steadily increasing volume from 1986 to 2001, advertising and promotion, improved quality, and more aggressive "supply control," greatly expanded retail advertising in the spring and summer. Most of this activity was caused by the inherent popularity of the strawberry and its contribution to chain margins. *What role does the CSC, or any commodity board, have in the competitive retail world with strong private sector influence, aggressively supplying and promoting their product?* The role of the CSC's incentive payments and private retail programs is detailed in Chapter 6 on variables. By 2001, all of the above factors contributed to 4,220 ads by 278 retailers in the top 50 markets, tracked by Leemis. By comparison, 3,234 ads by 440 retailers were counted in the same markets in 1991. During that time the frequency with which retailers were running strawberry ads had increased from 4.4 to 8.9 per season. Strawberries represented 5% to 6% of gross produce department sales and were among the top three items in the produce department throughout the April–May peak. The number of ads, frequency, and strawberry percentage of gross produce department sales, are barometers for measuring the possible demand effects of CSC and private sector programs and are analyzed in Chapter 6 on variables.

Trade advertising and public relations maintain visibility for California strawberries, which growers feel is necessary. One espoused raison d'etre for these actions is "maintaining a presence" because all other commodities are engaged in similar programs. While most of the 1990s activity was a continuation of programs established in the 1970s–80s, except with larger budgets, the CSC emphasized that the future of advertising and promotion programs should be based on category management data and the ad incentive would program would continue to be the "work horse" of the marketing program.

Notwithstanding the 34% decline in the number of tracked retail accounts from 444 in 1991 to 290 in 1999, the total number of ads only slightly declined from 2,181 in 1991 to 2,051 in 1999. Summer ads have actually increased 38% to coincide with the increasing summer volume and spring ads dropped less then 6%. This represents a significant increase in the number of ads run by each retail unit. In 1991, the average chain ran about 7 strawberry ads during the course of the entire season, while in 1999, the average retailer ran over 7 ads during the peak season alone, and 11 ads for the entire season.

Table 8
Fresh Strawberry Advertising History: 1991–2001

Year	# Retail Chains	Apr-Jun Ads	RWV	Avg Ads	Jul-Sep Ads	RWV	Avg. Ads	Total Ads*	RWV	Avg. Ads
2001	267	2,292	1,868	8.6	917	765	3.4	3,689	3,034.30	13.8
2000	278	2,481	2,016	8.9	1,248	922.38	4.4	4,270	3,385.48	15.3
1999	290	2,053	1,634	7.1	898	648.93	3.1	3,529	2,735.79	12.2
1998	301	1,935	1,501	6.4	912	701.91	3.0	3,287	2,545.50	10.9
1997	307	2,160	1,605	7.0	739	497.02	2.4	3,628	2,672.06	11.8
1996	319	2,107	1,486	6.6	821	522.31	2.6	3,528	2,497.80	11.1
1995	320	1,921	1,423	6.0	880	610.13	2.8	3,414	2,498.42	10.7
1994	320	1,860	1,321	5.8	875	619.2	2.7	3,403	2,432.23	10.6
1993	409	1,940	1,238	4.7	1,060	551.71	2.6	3,408	2,209.20	8.3
1992	427	1,941	1,229	4.5	528	289.69	1.2	3,000	1,840.04	7.0
1991	440	2,181	1,384	5.0	548	275.44	1.2	3,234	1,969.67	7.4

Source: CSC citing Leemis Market Research
RWV=Raw Weighted Value

An evaluation of the major trends in the retail, food service, and producer sectors is necessary in order to provide further background, a better understanding of the strawberry market, and how the CSC and grower/shipper function together. Produce represents $83 billion, or 11% of the food industry. Currently, the top 20 retailers account for over half of U.S groceries sales. Sales of the top 4 chains rose from a 16% share of total grocery sales in 1992 to 29% in 1999. The top 10 integrated wholesalers/retailers, including voluntary wholesalers like Supervalu, accounted for 50% of 1999 grocery sales.

Food retailing contributed 52% and Food service 48% of the total US food system of $788 billion. Supply centers like Wal-mart, which made up 74 % of retail food sales with $70 billion in 1999, were expected to hit $180 billion by 2003. They are contracting with preferred suppliers and implementing new technologies such as vendor managed *automatic inventory replenishment*, which is further reducing the spot market. This market power is reflected in the new demand for fees and services. The retail interest in incentive allowance for ads, is a reflection of the market power, and causes one to ask, "*Would they advertise without industry contributions?*" Although the answer to this question is probably 'yes,' the CSC just in 2004 has discontinued this expenditure *and shifted its focus and budget, to public relations and scientific studies on health aspects of strawberries.*

The proposed model, including barometers for measuring the effect of board policies on demand shift variables such as prices, quality, advertising/promotion, and consumer attitudes and behavior, should enable industries to better evaluate the effect of the budget size and scope of intended future board policies on these variables and demand.

The U.S. consumption of fresh produce has gradually increased from 287 lb per year to 322 lb during 1990–1999, while fruit consumption has also moved from 114 lb to 132 lb. Produce sales by departments have increased from 9.58% in 1990 to 9.70 in 2000. The spring-strawberry category contribution to total produce dollars increased from 4.5% in 1998 to 6.1% in 1999, and decreased from 5.5% in 2000 to 4.9% in 2001. The summer-strawberry percentage for the same periods ranged from 2.9% to 3.3%, 3.8%, and 3.8%. The total annual percentage contribution is 3.4%, 4.4%, 4.3%, and 4.4%, respectively. *Weather, supplies, and price can effect both spring and summer category contributions.* The top fruit categories contribute almost 40% of total department sales. Specifically, bananas apples, grapes, citrus, berries, and tree fruit, at 8.4%, 7.3%, 6.9%, 5.8%, 4.9%, and 4.8%, make up the 40%. Proprietary and day neutral varieties have extended the season through the summer and fall. They provide the total season framework for analyzing the strawberry market in five specific stages: Stage I—Beginning of the season to Easter, Stage II—Easter to Mother's Day, Stage III—Mother's Day to July 4, Stage IV—July 4 to Labor Day, and Stage V—Labor Day to the end of the season.

This *five season framework* will be utilized to evaluate past board marketing programs by incorporating them into the descriptive model widely used in analyzing agriculture and commodity systems in order to *study the relationship between the demand for fresh strawberries and the variables affecting demand.* Again, the central question to be addressed is whether the CSC mandatory check-off programs or assessments have caused demand to increase and whether that increase has had a positive effect on the producers of strawberries. The information developed from my descriptive model will be applied to understand the strawberry market and to describe the basic features and internal functioning of the system. Hopefully, the model may be helpful for all commodity boards that desire to measure/evaluate prior actions and project the effects of proposed programs on retail, food service, consumer demand, and ultimately the effect on grower welfare. Since the effect of generic consumer advertising is the subject of most analysis and measurement in the economic literature, the insight presented by this literature is helpful in understanding past studies of marketing orders, and are relevant to understanding the CSC and the effectiveness of prior advertising programs. Additionally, the economic studies provide the framework for the development of barometers for at least some measuring of the effect of board programs designed to increase demand.

Chapter 4

Literature Review of Generic Commodity Advertising and Promotion

The CSC advertising program from 1990–2002 consisted mostly of retail promotion activity without direct generic consumer advertising and is more readily evaluated in terms of changes in sales, prices, total ads, market share, and the list of barometers, Chapter 5. Generic advertising, on the other hand, is much more difficult to evaluate because of its broader reach and the fact that many other uncontrollable factors, such as price, volume, substitutes, cross-advertising, and so forth, may influence sales (Brader). The smaller, federal marketing order programs tend to place more emphasis on promotion programs than on generic advertising, while larger programs involving such commodities as fluid milk, frozen orange juice concentrate, livestock, cheese, cotton, wool, and pork, expend millions of dollars on consumer advertising. Most of the literature involves analysis of industries with supply restrictions and fails to distinguish between generic advertising and promotion. Generic advertising is paid media advertising, while promotion may involve in-store promotion, couponing, special allowances, demonstrations, and incentive ads. Brader makes this distinction, without differentiating in his analysis of federal marketing orders, causing difficulty in evaluating the effectiveness of either promoting or advertising.

Most of the literature involves the econometric analysis of the effects of generic advertising, not promotion, on sales and does not include market analysis currently suggested by most economists (Johnson) and the writer. The reason for including market analysis is the need for *some tangible asset, other than sales and advertising expenditures*, the traditional indicator of advertising effort, for measuring the effects of advertising. Blaylock would be reluctant to solely rely on empirical results as sufficient criteria for evaluation of an advertising program. Many other economists in this generic advertising field are uncertain about the effects of generic advertising and question the data available for analysis (Kinnucan). Inadequate data used in generic advertising research, which incorporates advertising in a demand system and determines how to measure cross advertising effects, has made it difficult to determine if advertising pays (Hayes).

The Giannini Foundation publications on grapes in 1997 and prunes in 1998 are the most analytical and provocative for the unregulated produce industry without supply control. These studies introduce *demand shift variables* while utilizing econometric and market analysis. Measuring the advertising effect on demand is a unique problem because, as stated by Alston and Chalfant, "The effects we are trying to measure are small relative to the effects of prices and other things that changes demand. The problems are especially pronounced, because the effects of advertising are complex, poorly understood, and small." Advertising and promotion are the areas most subject to industry and control, although other variables are more significant. The econometric analysis, which predicts increased demand and price because of the positive advertising variable, *fails to differentiate between generic advertising and promotion*, the importance of different media levels and types, and excludes private shipper and chain retail advertising. The positive advertising effect on demand is simulated, *while their data indicates that demand and prices are not higher and are actually lower in some years, raising the question of "reasonableness."* The grape simulation uses commodity board data on advertising expenditures and relates it to augmented sales and demand, but does not consider private and chain sector advertising, and thus overstates board effect on demand. The Giannini prune study also suggests a positive demand/price effect from TV ads, but could not draw any conclusions about causal relations because of "confounding influences" that occurred during the test by Nielson. The TV ads, assumed to increase demand/price, coincided with increases in sales price and promotional "deals" by shippers, which in effect reduced prices. Increased demand is not directly related to advertising or accurately measured because of numerous price variables, weather factors, and unique merchandising practices, such as reduced pricing before an advertising or promotional activity (McClelland).

There appears to be consensus among economists that *market analysis is necessary* to understand the variables that effect demand and have improved integration of research methods from marketing and economics. There is neither clear nor convincing evidence on the magnitudes of shifts in demand brought about by advertising, how demand is shifted by advertising and promotion, or the incidence of the benefits from such shifts. Supply responses, cross-commodity effects, and wearout of repeated information to retailers and the public, are among the factors complicating evaluation research (Johnson). *The barometers, or indicators of an advertising variable effect on demand,* are currently included in economists' models and require some sort of subjective evaluation, emphasized and expanded upon in my model (See Chapter 6). Market analysis incorporated into econometric models as an *ad effectiveness variable* progressed from the traditional view that advertising expenditures and increase in product sales are the sole indicators of promotion effort (Blaylock).

My descriptive model substitutes marketing information, specialized data, and some controlled experiments for the prior structure used in the economic approaches for assessing the impacts of advertising (Johnson). Current commodity advertising and promotion literature represent an integration of concepts from economics and marketing but does not analyze the effect of promotion as a separate and distinct demand shift variable. The Giannini grape study indicates that 90% of the grape budget is marketing, 50% is consumer advertising, and 25% is merchandising-promotion, but does not analyze the effect of retail advertising (number of ads) on demand, price or awareness and attitude data collected from opinion surveys often used as a guide to advertising effectiveness. The Economic Research Service (ERS) of the USDA identified changes in prices, incomes, and donations of dairy products as having the largest impacts on consumption but recognized that other factors such as advertising were important. The ERS also recognized the difficulty in assessing the impacts of brand advertising from generic and *the importance of utilizing data from household panels, supermarkets and food service establishments. Again, integrating research methods from economics and marketing are emphasized while the distinction between advertising and promotion is lacking (Brader). Data collection and reliability remains a basic difficulty in measuring or assessing the effect of advertising or promotion, such as, securing privileged information from commodity boards that could indicate negative aspects of their respective programs.* As Kaiser indicates, "It is a fact that the quality of the available data is sometimes insufficient to obtain the desired stable estimates." The previous literature noted above, has provided me with the framework for this study and the model we hope will help evaluate commodity board programs.

CHAPTER 5

Descriptive Model for Measurement and Effectiveness of Board Policy

The two central questions addressed by this study are whether commodity board "mandatory check off programs," or assessments, have caused demand and supply to increase and whether the increases have resulted in *measurable performance outcomes*. These *outcomes* include growers' price, farm value, and proxies for demand, such as attitude changes, strawberry market share at the retail and food service level, sales dollar contribution per store, strawberry category contribution to total produce dollars, and the number of retail ads. Growers' costs, total production and production per acre, and profitability, as affected by CSC activities in pomology and horticulture, are the *measurable performance supply outcomes*, evaluated in the supply variable section, Chapter 6.

"Models are generally constructed to provide information about systems. Depending upon research objectives, models have one or more of the following orientations: descriptive, explanatory, predictive decision-making," states Rauser. Descriptive models have been widely used in analyzing agricultural and commodity systems. A practical, decision making model, which requires forecasting and predicting capabilities, may not be effective if the markets being studied are not understood in terms of their relationships. *Similarly, a useful model for a member of a commodity board may be different from a useful model for an academician.* The research community should keep in mind the real potential consequences in contrast to theoretical implication. The information developed from my descriptive model will be used to understand the strawberry market and to describe the basis features and internal functioning of the system. The following model will be used to explain the relationship between the variables and to indicate the functioning of the CSC as far as its relationship to grower/shipper members, their separate and joint affect on demand and price, supply and grower costs, and measurable performance outcomes.

In order to analyze the relationship between the quantities demanded and the variables affecting demand, the standard theory of demand suggests a model in which the Quantity demanded, Qt, depends on the corresponding Price of Strawberries, PS; Quality Q; Frozen strawberry Volume, Price, and Inventory, FVPI; the Price of other Substitute Fruits, PSF; Total income or EXPenditures on all goods, EXP; Other, OTH:

$$Qt = f\,(PS,\ Q,\ FVPI,\ PSF,\ EXP,\ OTH)$$

We will assume that EXP is constant and although constant tastes and preferences for strawberries are implicitly assumed, marketing study data will be introduced which will indicate the CSC has and continues to affect the Other variable.

In order to accommodate changes in preferences arising from promotion or anything else that may affect demand, the model is augmented with additional demand shift variables, such as promotion/advertising by the CSC, individual shipper/producers, and retailers. The augmented demand model will include the following demand shift variables: Price of Strawberries, PS; Quality, color, and new varieties, Q; Generic advertising/promotion, CSCPROMO; advertising/promotion by chain stores with/or without grower/shipper support, PRPROMO; Committed and or Contract Pricing between shipper/grower, and retailers, CP; Freezer Volume, Price, and Inventory, FVPI; the Price of other Substitute Fruits, PSF; Total income or EXPenditures on all goods, EXP; Increased consumer health consciousness and interest in natural food, demographic changes, more meals eaten away from home, the longer season for strawberry varieties in other localities, and imports from China, Latin America, and Mexico, OTH. Incorporating the demand shift variables leads to the augmented model:

$$Qt = f\,(PS,\ Q,\ CSPROMO,\ PRPROMO,\ CP,\ FVPI,\ PSF,\ EXP,\ OTH)$$

In order to assess the effects of the variables on demand, supply, and producer's welfare, there is need for some tangible asset to justify growers paying the bills for marketing order programs. The tangible asset typically used has been advertising expenditures as the sole indicator of promotion effort. The barometer of a program success has been product sales attributed to advertising expenditures

(Blaylock). These barometers are inadequate to answer the question, *"Was the advertising effective and did it pay?"* Advertising expenditures do not equal effectiveness, and measuring advertising effectiveness is more difficult because it is not targeted as is promotion. The effectiveness of promotion effort can be readily evaluated in terms of changes in sales because, while other factors may have some impact, the promotion is often very specifically targeted in a known, and somewhat controlled, marketing environment (Brader). Sales data alone, which is affected by uncontrolled, interacting factors, such as volume, price, quality, competition, and substitutes, is a necessary but not sufficient, barometer of measuring generic promotion activity. Currently, both the CSC and private sector promotion and category management programs are essentially the only advertising expenditure (no media), and thus the most relevant, although the barometers will be the source of information to measure all variable effectiveness, specifically public relations and research in the health benefits of nutrition, OTH. The supply side of the model, which has been and continues to be the main variable affecting price, is the most affected by CSC expenditures. The budgets for University of California research in pomology and horticulture have and continue to be the driving force for increases in yields, reduced costs, and acreage/production expansion. The following model will provide the framework for evaluating the effect of the CSC on supply and price, using the barometers for establishing a tangible asset for effectiveness of CSC or other board programs. The marketing board adoption of this model is necessary for effective grower accountability, much the same as a CPA assurance of proper accounting procedures:

S = f (A, P, YPA, CPA, VAR, FVIP) <u>or Acreage, Production, Yield Per Acre, Cost Per Acre, VARieties, Freezer Volume Inventory and Price</u>

A partial list of barometers will also be applied to prior CSC programs in order to evaluate their effectiveness, although all of the information essential to the use of any specific barometer will be left to future researchers. This *framework* will hopefully be provided for future commodity boards of directors to evaluate past and future programs utilizing the above model, and information from the following barometers for assessing measurable performance outcomes:

1. Average nominal and real prices of strawberries
2. Net farm income
3. Price volatility

4. Total sales
5. Total production and production per acre
6. Production cost per acre
7. Gross and net returns per acre
8. Per capita fruit consumption
9. Actual newspaper ads over time
10. Strawberry market share compared to other fruit
11. Measurable attitudinal and behavioral changes (*Proxies for Sales*), such as product awareness, ad and product recall, and consumer attitudes toward strawberries (see Achabal study, Chapter 6).
12. Strawberry sales dollar contribution per store.
13. Variety and quality changes
14. International strawberry fresh and freezer competition
15. Chain store concentration
16. Changes in packaging and multi-packaging
17. Advertising and promotion wearout
18. Food service total and relative growth
19. Fresh and freezer producer concentration

The major demand determinant is price. According to the Law of Demand, price, PS, is inversely related to quantity demanded, Qt, which is frequently explained by other variables in the system. Historical data clearly indicates that weekly, monthly, and annual market volumes explain a major portion of the variation in FOB prices. The remaining variables, some of which we can only partially evaluate because of limited data, must be explored by future researchers. *Complete evaluations of the variables that effect demand, requires industry interviews and the use of information secured from analyzing the barometers.*

CHAPTER 6

Evaluation of Model Variables Utilizing Barometers for Measurement of Board Policy

Demand Shift Variable: PS, Price

The market volume explains a major portion of the variation in FOB prices for fresh and processed berries. Tables 9 through 15 show volume and prices from 1988 to 1998. FOB prices tend to follow an annual pattern, falling when supply increases. Seasonal variations are evident as volume increases in January through May, when prices decline dramatically and rise in a similarly drastic manner late in the season as production phases out. The uncontrollable, strong seasonal characteristics of supply cause most marketing problems. The simple correlation for the two series is -53 indicating that, even in weekly data, the volume available in the market explains a major portion of the variation in the FOB price (Han). *All marketing boards should understand this relationship before embarking on programs to influence demand. Higher or lower prices do not affect supply since strawberry supply is not a function of price, only acreage, production, and yields per acre.* Under certain circumstances freezer prices may affect fresh supplies, as discussed in the supply variable section, (FVPI) later in this chapter.

The evaluation of demand enhancing programs also requires an understanding of price/volume relationships changing over the course of the season and the differing effects of demand shift variables on price. Our discussion of CSCQ, CSCPROMO, Q, PRPROMO, and CP will relate these variables to each of 5 seasonal market periods, analyzing and measuring their effects on the relevant barometers mentioned above.

Volume weighted nominal prices declined for both fresh and processed berries from 1988 to 1993, while real prices followed a more pronounced downward trend. From 1993 to 1998, nominal prices rose steadily and real prices less pronounced. Price/volume relationships change over the course of the season and these differences are important in understanding the effect of any demand shift variable on price.

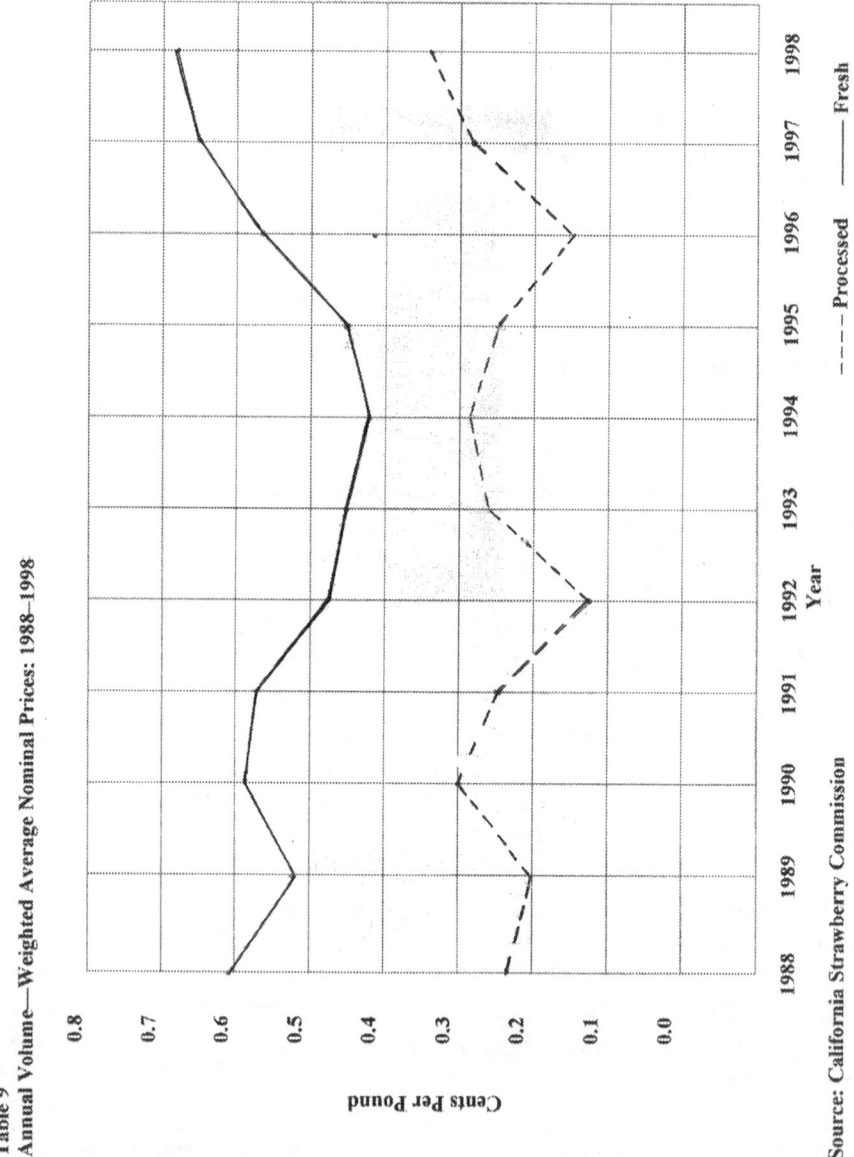

Table 9
Annual Volume—Weighted Average Nominal Prices: 1988–1998

Source: California Strawberry Commission

Table 10
Annual Volume—Real Prices: 1988–1998

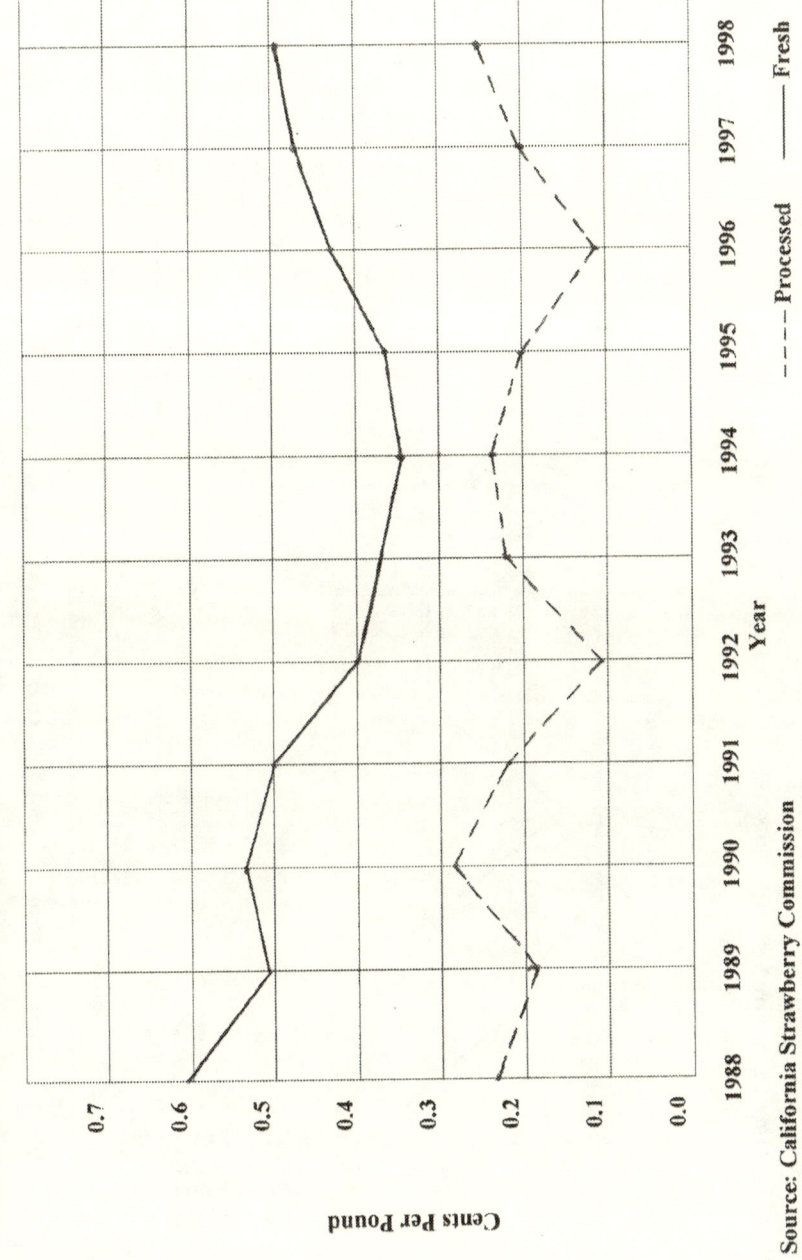

Source: California Strawberry Commission

- - - - Processed   ——— Fresh

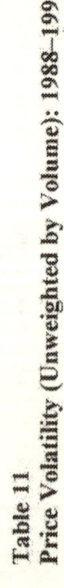

Table 11
Price Volatility (Unweighted by Volume): 1988–1998

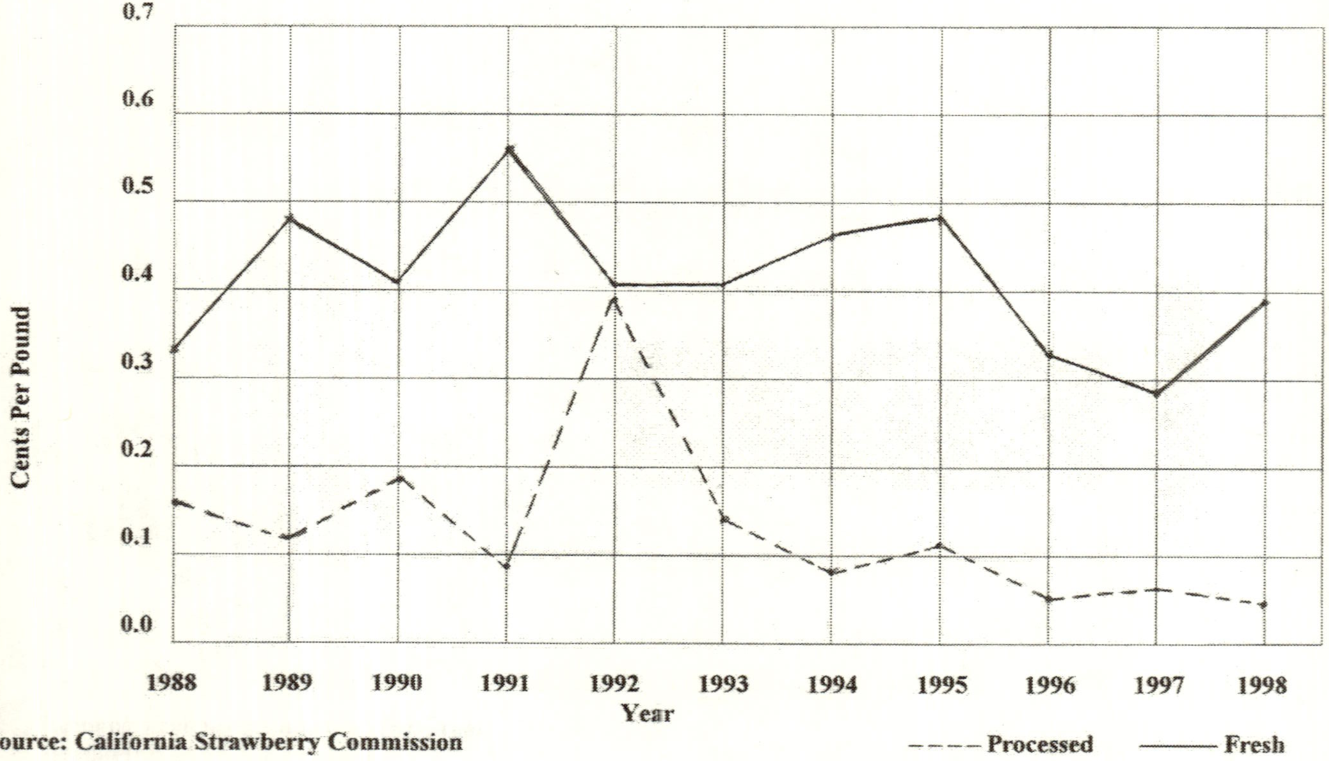

Source: California Strawberry Commission

Table 13
Season by Season Plots of Weekly Real FOB Price and Shipped Volume: 1991–1993

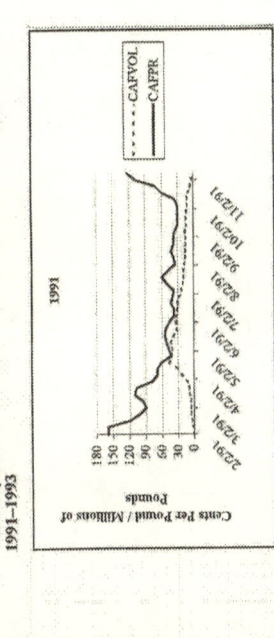

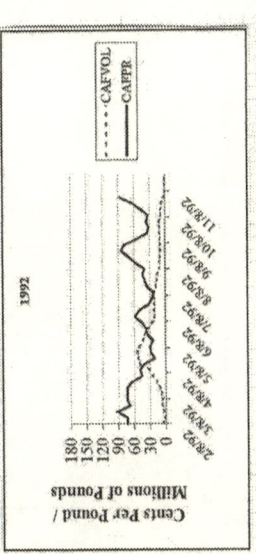

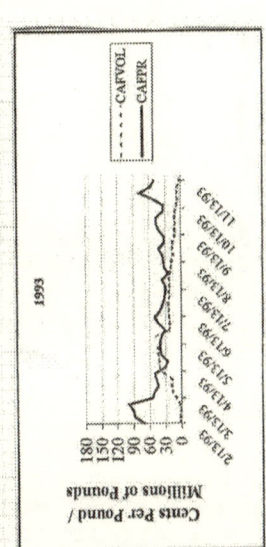

Table 12
Season by Season Plots of Weekly Real FOB Price and Shipped Volume: 1988–1990

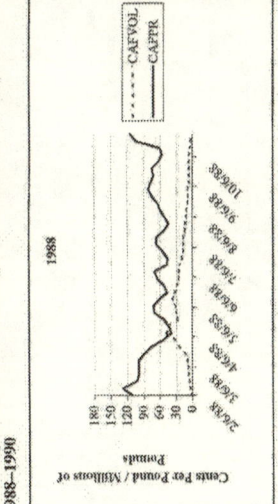

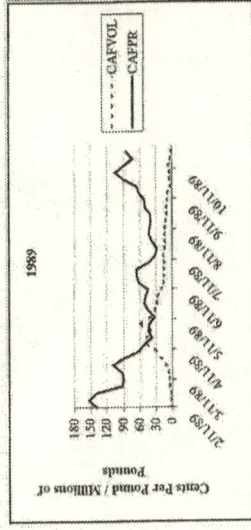

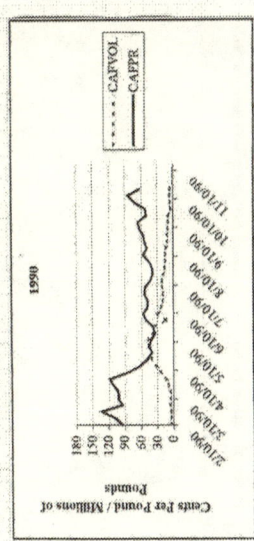

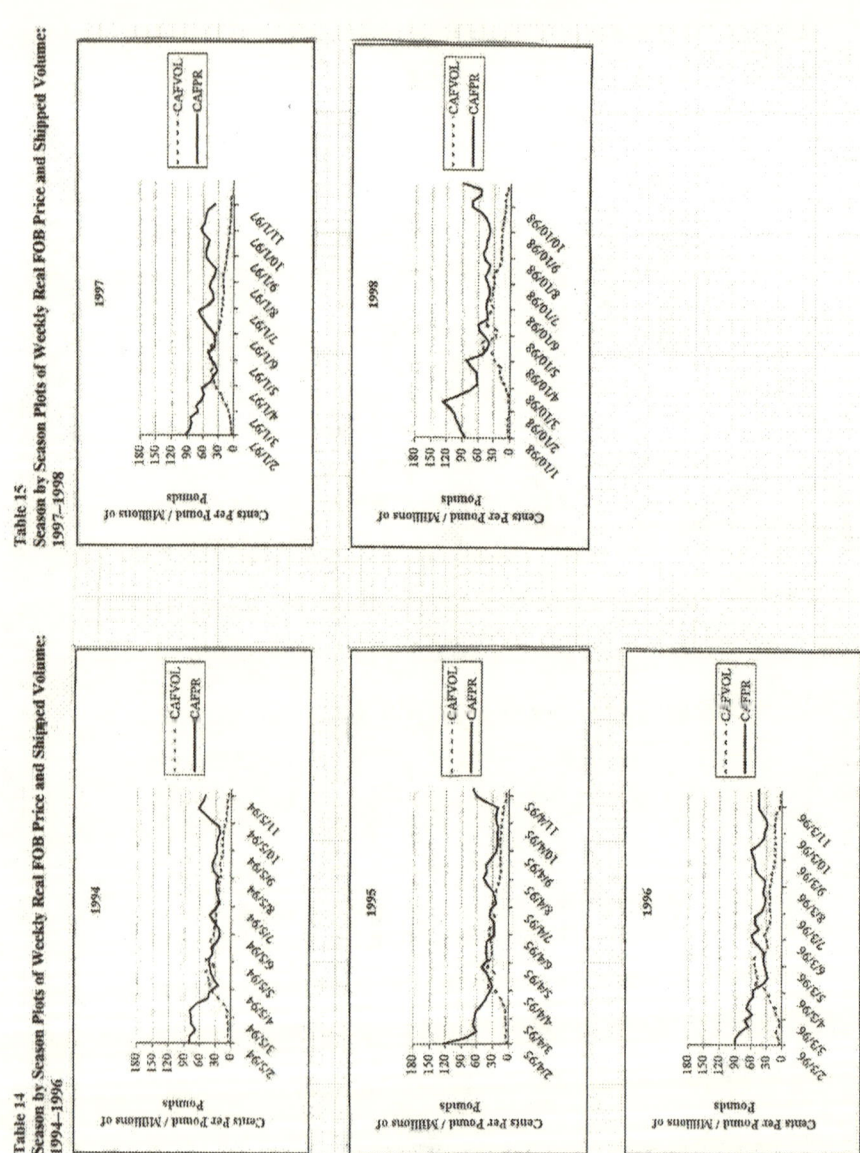

Table 15
Season by Season Plots of Weekly Real FOB Price and Shipped Volume:
1997–1998

Table 14
Season by Season Plots of Weekly Real FOB Price and Shipped Volume:
1994–1996

Tables 9 through 15 show that FOB price variability appears to have decreased over the 1988–1998 period. The available data, primarily from private sources, indicates increased price stability from 1958–1988, as sales planning rather than spot market dependence, became a more utilized marketing tool. The data indicates that the annual price variation has decreased. The coefficient of variation of annual average prices shows a slight but definite downward trend from 1988–1998 (Carter). Both real FOB prices and the range of variation have declined. Although Carter recognized the decline in weekly responsiveness to volume changes, he suggests the underlying economic relationships governing strawberries are moving toward a pattern of less volatile prices. He does not discuss these underlying relationships, although there is interest in future research into price volatility, pre-commitments and contract pricing. Also not mentioned is the role of the CSC or the private sector and their relationship to advertising and promotion. A major objective of this study is to describe and analyze this relationship, and the effect of the variables, CSCPROMO, Q, PSPROMO and CP on demand and price.

Although FOB prices generally do fall with increases in volume, there are examples of price declines with volume decreases. This can easily occur for systemic reasons. There are situations where *forward pricing*, based on anticipated volume increases that fail to materialize. The result is lower price and lower volume. Even though volume declines, prices may decline even further, if promotion prices are not projected accurately, or if forward pricing in not low enough to clear the market. A strong and statistically significant inverse relationship between volume and price, with constant variations make model prediction extremely difficult. For example, the price/volume data for May 2001 indicates a scenario of higher prices, crop value, and increased volume. Because of a late season and lost sales, and a delayed summer fruit deal, chains aggressively advertised increased volume at higher prices throughout May and June, and maintained strawberries in a primary retail store position. Demand increases because of this practice. Summer fruits are a demand shift variable and their shortage is an example of a positive effect on strawberry demand and price. With the exception of Florida citrus, the coefficients for all proposed substitutes for fresh strawberries were shown to have statistically insignificant effect on weekly FOB prices (Carter). However, the inability of the Carter model to predict demand variations in this situation, because of inaccurate, supply and demand predictions, for strawberries and substitutes, explain the difficulty of model predictability and reluctance of the CSC to rely on them.

The five distinct strawberry market stages, discussed earlier in Chapter 3, have been present since the 1960s. In Stage I, or the beginning of the season to Easter, USDA and proprietary data indicated higher prices in January, February, and March, with production from Southern California, Baja California, and Florida. As Baja and Florida utilize California varieties and horticultural practices, competition intensified and prices tended toward lower levels. However, weather in all areas was the determining price factor while disaster in one area was another's good fortune. Chile and other South American countries are shipping summer fruit during the winter-spring California season, frequently securing primary aisle space in retail stores. Per acre yields were increasing with the new hybrids, enabling the grower to reduce his costs per acre as prices decreased. In this stage, farm value is at its highest. Since 1983, Stage I prices were $11–$14 with March averaging $10, while during the peak volume months of April and May prices lowered to $5 or $6. Few substitutes are available before Easter under the spring market circumstances and demand elasticity has its lowest value in absolute terms. The demand will become more elastic later in the season as FOB prices are most responsive to changes in the amount shipped.

Very little CSC or private sector promotional activity takes place during Stage I with the exception of retailer contracts at market price in effect before Easter and fixed seasonal contracts with food service operators. The average shipper allocates 25% or less of his pre-Easter volume commitments during Stage I and generally at the spot market price. The CSC is heavily involved in public relations with retailers, foodservice, and industrial users. Prices are most sensitive, because of low supply and high demand.

Stages II and III include the periods from Easter to Mother's Day and Mother's Day to July 4th. If Easter is early, the period between the holidays is obviously longer. Weekly prices are very responses to volume shipped during this period, as well as Stage I. Historically, pre-Easter demand exceeds supply regardless if it is in March or April. Between Easter and Mother's Day, there are sporadic promotions and few substitute fruits are available. Chain *precommitments* at lid prices, two to three weeks in advance of Easter, are common and represent 33% to 50% of total volume for smaller shippers and 65% to 75% for larger shippers. Between Easter and July 4th, a minimum of 75% of total volume is precommitted at advance prices. These lid price commitments are subject to downward market fluctuations, not upward. Many food service contracts have fixed prices, with *Act of God* clauses. During these Stage 2 and 3 periods, April, May and June represented 54% to 58% of total fresh volume from 1990–2001 and 60% of total volume in 1980s.

The data on advertising is essential given the relationship between advertising and price. In 2001, strawberries were advertised 3,689 times between March and September in 267 chain stores nationwide. The average chain advertised almost 14 times as compared to 7 times in 1991 (See Table 8). Advertising decreased 14% from 2000, the result of a 7% decrease in strawberry volume. Because of adverse weather in 2001, volume disruptions leading into Easter caused retailer reluctance to advertise after the holidays. Summer advertising declined because production was down 18% to 21% over the previous year.

Advertising and promotion, as well as other variables in our model, are most important during Stages II and III, which represents a large percentage of the total volume and farm value. Again, the volume available in the market explains a *major* portion of the variation in the FOB price (Chalfant). Demand is the most elastic in Stage III, the peak production period, with an estimated price elasticity of—2.91 (Carter). With no substitutes during Stages I and II, demand is less elastic and comparable to Stages IV and V. Stages II and III are the most active periods for the CSC and for grower/shippers in their attempt to shift the demand curve with promotional effort directed to retailers and food service operators.

The summer volume periods of Stages IV and V, or July 4 to Labor Day and Labor Day to the end of the season, have increased significantly over the past 10 years (CSC Pink Sheets). The day neutral varieties, grown mostly in Northern California and Santa Maria, continue producing until November, or until frost or rain sufficiently halt commercial fruit production.

The California summer fruit season, including local fruits throughout the U.S, and the greatly expanded frigo plantings in Southern California, has provided a new marketing challenge. Prices during July from 1983 to 1998 were lower for 10 of the 15 years and for 12 of 15 years in August. Summer volume averaged 10 million trays in the 1980s, 12 million the 1990s, and 18 million in 2001. Retailers, who had been promoting strawberries and featuring aisle space for many months, were anxious to substitute other fruits from July 4 through Labor Day as the demand for strawberries decreased. Other commodity boards and grower/shippers of all fruits were advertising and promoting with the same techniques for retail space and food service attention. *Price and profit maximization will determine what produce items gain the retailer's order and display space.*

Stage V, representing the period from Labor Day until the end of the season, is highlighted by reduced volume and prices. Substitute fruits are available and there is low retail interest in advertising and promoting strawberries. Most warehouse stores, such as Costco, discontinued their use completely. During this Stage, volume doubled from 2 million trays in the 1980s to over 4 million in 1990 and doubled again for the next decade to 8 to 10 million trays in 2001. Prices have been lower in 9 of 14 years, in both September and October from 1988 to 2001. The development of day neutral varieties in Southern California (Oxnard) increased supply, which depressed a traditionally high priced market in late fall and reduced the average price for growers. During the months of September, October, and November, the average price was lower for the period 1993–1997 than for 1987–1992. Under these circumstances, a commodity board is tempted to interfere in the market by initiating programs to affect the other demand shift variables, *since suppliers can only directly influence price.*

Suppliers in there own self-interest appreciate that lower prices and increased demand are not an option unless a reduced forward price, less than the current spot market price, will result in increased demand. It is assumed by this decision that the current spot market price is too high for the anticipated supply increase and it may cause price to go lower than necessary to eventually clear the market, if sustained. "Upside price management" and contract pricing, discussed in the section on PSPROMO—CP, are partial solutions for market gluts caused by inaccurate market information.

Demand Shift Variable: CSCQ, CSC-University of California Research Model on Quality

The price and demand data are overstated, unless it is assumed that all varietal quality is equally satisfactory, during all five marketing Stages. All USDA price data is based on a "mostly" price and does not consider quality at shipping point or at destination. All varieties in Stages I and II are relatively good and acceptable to chains and consumers at competitive prices. In Stages III, IV and V, a price *range* develops based on proprietary and university varieties. As indicated in the section on pomology, the proprietary varieties usually command higher prices and are in the *higher* range of the *mostly* FOB prices, indicating a greater demand for superior quality, including color, taste, and longer shelf life.

The Quality, Q, demand shift variable ranks first after price in effecting the total demand for strawberries. It will affect prices mostly in Stages IV and V as warm weather in producing areas and destination markets reduce shelf life and appearance. The CSC model, which utilizes UCD for all research in horticulture and pomology, has been the successful vehicle for improving quality and increasing yield. Quality and taste have been the most important elements in consumer decision making when choosing alternative berry varieties, frequently reflected in proprietary brand preference. This was true in Stages III, IV and V, until recently, when a new variety developed by the university, the Diamante, provided industry competition for proprietary varieties. The taste and color are still inferior to proprietary varieties, and are disliked by processors because of white interior color. However, the firmness and tremendous per acre yields have overcome some of the competitive taste disadvantage and the long period of trade dissatisfaction with non-proprietary varieties.

Price differentials can be significant between university and proprietary varieties, especially in Stages III, IV and V, and proof of this is the increase in proprietary acreage in all districts of California, as noted in the section on pomology.

The gap appears to have closed somewhat, however, the decrease in independent growers using university varieties is further proof that demand and pricing favors proprietary varieties and organizations. From 2001 until 2004 planting, proprietary acreage increased from 29% to 38% of total acreage. During the same period in Oxnard, proprietary acreage increased from 13% to 21%, Santa Maria 2% to 3%, and Salinas-Watsonville from 13% to 14% of the states total. The success of proprietary varieties due to quality and higher prices is even more pronounced when we look at the percentage increase in acreage by district. In the three major districts, Oxnard, Santa Maria, and Salinas-Watsonville, proprietary varieties represent 59%, 15%, and 37% for 2005 planting, compared to 45%, 14%, and 31%. The proprietary varieties and horticultural progress have caused a relative shift in *their* demand curve with relatively higher prices than university growers and shippers. The clear differentiation between proprietary and university variety growers is now blurred as more growers and shippers combine private and public varieties. Organic acreage rose from 300 acres in 2001 to 541 in 2005 with the proprietary share at 50% (See Table 5).

These changes are exacerbated by the inflationary increase in production costs. All growers share a similar level of costs in each of the districts, whether proprietary or university growers. Therefore, net grower returns will depend on production per acre and price. The following Table 16 is a summary of approximate, average FOB prices, net returns per tray, and total cost per tray for all 3 districts, from 2001–2003:

Table 16
Grower Costs and Prices: 2001–2003

	FOB	Net Price (less crate & cooling)	Total Cost Per Tray
Oxnard	$8.78	$7.26	$6.50–7.00
Watsonville & Salinas	7.61	6.25	6.00–6.80
Santa Maria	7.14	5.87	5.50–6.00

Source: *Sample Costs to Produce Strawberries*, by the University of California Cooperative Extension. (2004) South Coast Region (Oxnard): Yields and Returns, Table C, p.5; Net Returns Per Acre Above Total Cost, Table 4, p.17; South Coast Region (Santa Maria): Yields and Returns, Table C, p.6; Net Returns Per Acre Above Total Cost, Table 4, p.17Central Coast Region (Watsonville-Salinas): Yields and Returns, Table C, p.5; Net Returns Per Acre Above Total Cost, Table 4, p.15

Additionally, a range of the average tray price for each district, a range of yields per tray/acre, and net returns for each yield and price, provide a guide for net returns, based on a specific level of prices and yields. The data illustrates the importance of price, assuming some comparability of yields. The dominance of price and quality as demand shift variables is unquestionable and the interrelation between quality, demand and price, is crucial to understanding the importance of the CSC in developing research programs that improve Quality, Q. At least 5 private shipper-breeding programs are in existence, 3 of which are competitively superior to university programs, especially in Stages III, IV and V. The California Strawberry Commission Annual Production Research Report for 2003–2004 does not mention research effort to address the competitive problem of proprietary varieties undermining the university program, as acreage shifts to more profitable quality and production patterns. The CSC research budget for 2003–2004 allocates a total of approximately $1 million of a total $7 million, or 15%, compared to 25%–30% in most prior years. The effect of the Quality variable Q, on demand and price, has not and is not appropriately being measured. Some of the barometers listed above that should be used for this purpose include: Varietal acreage, price and yields per acre, industry interviews including growers, shippers, and retailers, and use of contract pricing for proprietary and university varieties. Much of this information may be difficult to obtain because of its proprietary nature, however, much is known and can be deduced from acreage shifts and price ranges provided by the USDA. My personal experience provides clear evidence from 1958 to 1999, that a $1 to $2 difference between university and proprietary varieties, because of quality, is not uncommon especially if we compare prices at destination markets. The USDA Federal Market Report reports ranges of prices, and generally report, "poor lower," meaning poor quality. At California shipping points, the USDA Fruit & Vegetable Market News daily report includes prices quoted by buyers and shippers and there is reluctance to describe inferior quality or to reveal ad prices or unsold product, which still will be sold in some market at below the mostly spot market price (See Appendix 2). Examples of Quality influence on demand and price occur in the daily USDA Fruit & Vegetable Market News report: The shipping point report for March 2, 2005, indicates "demand best fairly light, others light, market lower, wide range in quality, $10.90–$14.90, mostly $12.90; and for the Los Angeles market that date, "$14–$18, mostly $16–$18, fair condition $10–$12, few best $17–$18, one label higher;" Chicago for the same date, "Large to Xlarge $19–$20, Fair appearance $14–$16 and ordinary appearance $12–$14; and finally New York, "market lower, $18–$21, fine appearance $24, medium large $14–$16, few $13, fair condition $12." Recent industry interviews confirm my own experience with respect to the wide differential between proprietary and university varieties because of quality and color while taste has currently become the dominant consideration for large mainstream discounters. In prior years, some upscale chains were concerned about taste and color and would only purchase proprietary varieties. This shift has accelerated as evidenced by the proprietary acreage increase. One buyer expressed the taste interest with a request for information on the *brix* content and color variation, including percentage of green on the shoulders of the berry. Brix count has always been a factor in grading frozen berries but this is the first request that it be considered in the "specs" for a fresh berry. Bashas, an Arizona chain of over 100 supermarkets, features Driscoll at one dollar above a university variety, during a period of excessive supply, but also markets only Dricoll for much of the year. In May of 2005, Driscoll's were advertised, with a sample indicating their superior taste and priced at least one dollar above competitive chains.

The CSC and all commodity boards must annually analyze their budgets and priorities; the proposed model and barometers provide the tools for this task. Board accountability to the grower is a must. The assessment of the Quality variable has not been adequately measured, and therefore the competitive nature of proprietary and university growers and shippers has not been and currently is not addressed. The criteria required for a successful California strawberry cultivar are constantly evolving but the target traits outlined by Shaw and Larson in the recent production research report are *not* significantly different than in prior years. "The traits include: improved production attributes (marketable yield, production pattern, fruit size, ease of harvest), superior quality for fresh and processing markets (fruit appearance, color, flavor firmness, shipping quality, and shelf life), and resistance (tolerance) to pests, pathogens, and environmental factors or stresses. New cultivars must meet exacting minimum standards for all these traits but programs currently emphasize three target areas: Fruit quality, harvest efficiency, and biotic and abiotic tolerance."

Shaw, Larson, nor the CSC has addressed the problem of the competitive imbalance between them and proprietary pomologists. There is no doubt that the immediate goal of Shaw and Larson, as stated, is to release new cultivars *"that possess a combination of traits superior to its target predecessor, and with adaptation to the cultural treatments available at the time."* The Ventana is an improvement over the Camarosa in that it is earlier, larger, a heavy producer, and with a much lower cull rate which directly reduces labor costs. However, it is unlikely to produce significant amounts in May and June, thus providing no freezer crop and also seems particularly vulnerable to rain. Additionally, it does not have a good taste compared to proprietary varieties or Camarosa. An August 2005 Newsweek article, quotes a Whole Foods Market produce expert, "Look for a deep red, without white coloring around the stem. Top pick Camarosa variety, widely available and has a good flavor." Growers were planning a major shift from Camarosa to

Ventana, but the taste element surfaced in an important way, and Camarosa reemerged to cause an almost even proportion with Ventana. Whether there is a net gain for the university varieties remains to be seen, *but the question of relative taste and quality has not been adequately addressed* and the price differential between proprietary and university varieties in March of 2005 remains $1 to $3.

The day neutral varieties, Selva and Diamante, have been improvements in the ability of Santa Maria and Salinas-Watsonville university growers to compete with the proprietary varieties from June through November. The gap remains inordinately large because of quality, particularly the superior *taste and color* of proprietary varieties. Although the gap may have narrowed, the reality is that the proprietary market share for Stages III, IV and V has increased considerably because of *taste and color*. The largest proprietary acreage increase has been in Oxnard, which has experienced a 40% or 1,932 acre increase from 2004 to 2005. This increase also represents a 28% expansion in their district share and a 36% rise in the states percentage. (See Tables 2 through 5) Shaw, Larson, and the CSC have failed to address this competitive problem and have missed their goal with regard to new cultivars, previously quoted. The proprietary acreage total and relative increases are proof of the pomology and quality failure in achieving industry goals, especially a competitive playing field. The lack of cooperation between Shaw, Larson and all of the other university scientific cooperators has reached a critical level, according industry to interviews, and may be affecting horticultural and pomology advances.

The CSC and other boards should construct a basic research project on *quality* and the other *traits* described by Shaw and Larson. The problem in an industry with a CSC-university variety, used by all growers including those firms with proprietary varieties, is a political one. The questions posed to the entire industry, retailers and consumers, would pertain specifically to current varieties and help to provide an answer that has persisted since 1958, *Why has the university been unable to compete with proprietary research organizations for varieties producing during Stages III, IV and V, June through November?* The current question would pertain to Camarosa, Ventana, and Diamante, compared to Driscoll, Wellpict, and Nelson (an independent proprietary breeder) varieties. In 1999, I recommended outsourcing some portion of the CSC pomology budget to provide competition, which has been very effective in many areas of our society. The conclusions of the research project, which should be continuous and may not please everyone, would provide objective information on the vital issue. I would even go a step further and recommend that the research project should include varieties not yet released for nursery replication. Such information would provide a *taste and quality* review, much like the production, yield, disease resistance, etc, information provided by grower field experiments.

Illustration on the following pages:
University of California, Davis, and proprietary breeders.

Picture 17. Top: Victor Voth, Dr. Douglas Shaw
Proprietary Breeding: Top right, Tim Miyasaka. Center: Ken Morena, Gary Jertberg,
Clint Miller. Bottom: Bill Moncovich, Steven and Richard Nelson.

Demand Shift Variable: CSCPROMO, California Strawberry Commission Advertising and Promotion

The generic advertising/promotion variable has been measured in the literature with questionable success but its effect is relatively small after considering the effect of price on demand. Brader mentioned the distinction between promotion and advertising but did not develop it. Furthermore, he and Alston did not distinguish between board advertising/promotion and private sector advertising/promotion. The economic assumption that advertising increases demand is based on counter factual simulation, but even in this regard there are many negative factors influencing any simulation, namely, "cross commodity effects" and "leaking," to other markets (Johnson). Commodity boards seek "measurement of effectiveness" from economists, consultants, and staff. To do this the market place must be analyzed to determine how the advertising/promotion activity functions and then measure its quantity and effectiveness, using a wide range of barometers. Following comprehensive analysis, many board members have questioned and often reduced advertising and promotion budgets because, as they say, "It doesn't pay," or, "We don't have the funds to advertise effectively."

Most commodity boards, including the CSC, have a market research component, but not usually including econometric research, because of the need for some tangible asset to justify paying the bills. The tangible asset, typically used as the sole indicator of promotion effort, has been advertising expenditures, while product sales, attributable to advertising expenditures, have also been considered as the barometer of a programs success. These barometers are inadequate foe answering the question, *Was the advertising effective and did it pay?* The fact that advertising was made does not mean that it was effective, plus advertising effectiveness is more difficult because it is not targeted, like promotion (Brader). Sales data alone, which is affected by uncontrolled, interacting factors, such as volume, price, quality, competition, and substitutes, is a necessary, yet not sufficient, barometer of measuring generic promotion. The great number of things occurring in the marketplace and industry make estimation difficult because of the "degrees of freedom" problem encountered in econometric estimation (Tolley); therefore, the model proposed in Chapter 5 appears to be more tangible for commodity boards than the grape study's econometric model.

The continuous acreage and production growth of the strawberry industry has been a marketing challenge resulting in a gradual increase in the CSC assessment to a maximum of 5 cents per tray by 1981. This rate persisted until 2004, when it was lowered to 4.5 ¢, in order to reduce the surplus resulting from record acreage and production. Generic advertising/promotion encompasses a wide variety of techniques that are designed to apply upward pressure on demand by the consumer, retailer and food service sectors and to increase or stabilize price. Promotion, CSCPROMO, most often includes four main elements: consumer advertising, trade advertising, consumer public relations, and retail promotion/merchandising.

In 1974, the California Strawberry Advisory Board developed and began using the advertising approach to market its crop. Research showed that this first effort, consisting of low weight TV ads in 20 markets, increased demand for strawberries by encouraging retailers to display, feature and advertise strawberries more than in the past. The research indicated that the TV commercials had little or no effect on consumer awareness or purchase of fresh strawberries, but did tend to encourage retailers to display, feature, and advertise strawberries. This research offered insight into consumer behavior, number of ads, retailer interest in displays, use of point of sale material, quality, shelf life, cleanliness, lack of loose berries and degree of stains. All these barometers provided *a measurement of possible effects* of the advertising and merchandising effort by the CSC. An increase in demand was indicated by the fact that in 1974 the industry marketed a 26% larger crop than in 1973, but at an average FOB price 2% higher. Because of the volume and price barometers in this specific instance and with the expectation of increased production in 1975, the CSC board decided to turn to the low weight TV effort, this time in 44 markets, as a means of increasing demand. Although production was down in 1975 from 1974, price was up 8% and CSC efforts could have been the reason. Again, a production increase was projected for 1976 and an expansion into 58 national markets began. In addition, a lab-type test of heavy weight TV was conducted to find out if a more expensive consumer *pull marketing* program would be would be an effective way to increase demand and thereby market future increased production at higher prices. The expansion of this program, together with other merchandising and publicity efforts helped generate additional demand, which in turn, allowed the industry to market successfully larger crops without declines in average grower prices. This effect can be seen in the following Table:

Table 17
Fresh Strawberry Shipments, Average Price, and Crop Value: 1973–1977

	1973	1974	1975	1976	1977	vs76 %	vs75 %	vs74 %	vs73 %
Shipments	17,696	22,292	20,672	21,759	26,335	+21	+27	+18	+49
Price Per Tray	$3.59	$3.67	$3.67	$4.42	$4.54	+3	+14	+24	+26
Crop Value	63,554	81,769	82,355	96,147	96,147	+24	+45	+46	+88

With the low weight, advertising/merchandising *push marketing* approach nearing its maximum effectiveness in addition to indications that standard low weight TV programs had begun to wearout in terms of its ability to encourage retail cooperation, the board in 1976 tripled the TV advertising weight in two controlled test markets to discover if additional *pull* could be developed, while creating additional *push*. The CSC conducted the most sophisticated and extensive independent tests during 1976 by ADTEL, a division of Booz-Allen & Hamilton. It was a system for measuring sales differences that result from testing TV-advertising alternatives over a period of time. Market Science Associated analyzed ADTEL data and measured different weight levels or gross rating points (GRP) of California strawberry advertising in two isolated and typically demographic test markets. Two different test groups of 1,000 families each, in these two marketing areas, were exposed to varying levels of TV commercials of a large number of food products, including strawberries. They maintained weekly diaries from which data could be drawn. The conclusions were:

1. The TV ads had a significant positive effect of 36% and 25% on household purchases of both fresh and frozen strawberries in the two markets compared with families receiving the Board's regular level of advertising weight.
2. The increased sales effect of heavy advertising appeared almost immediately and was most evident during peak volume sales weeks for each market. This would be in Stages II and III of the marketing season.
3. More families, which had purchased strawberries only once every two or three weeks, were now purchasing almost every week. This accounted for the 25% to 35% higher purchase level.

4. The timing of advertising to the period of greatest strawberry availability in each market appears essential to maximizing CSC TV advertising effectiveness.

5. The springtime advertising also increases sales of frozen strawberries.

Armed with this data, the board decided to expand the pull marketing program, since it appeared to be a more effective and efficient way to market strawberries. The use of market information and applying available barometers enabled the board to carefully use existing research before venturing into expenditures with only guesses about possible outcomes. The proposed model should be useful for measuring the effectiveness of commodity programs involved with pull TV advertising, and push techniques including newspapers, magazines, tie-in promotions, merchandising, displays, and ad incentives.

In order to address the apparent wearout of low weight TV ads as an influence on retailer's cooperation, newspapers were tested in 1976 in 8 markets and 4 received newspapers alone. Newspapers were scheduled parallel to TV, with one 300 line insertion per week on best-food day for the same 6 weeks. The objective was the same, to encourage retail cooperation and the board audited retail newspaper features in the 8 markets, and 4 matched control markets.

The following Table 18 illustrates that newspapers were at least as effective as TV in their ability to generate retail features. Newspapers and TV combined were more effective but not enough to justify their cost. Newspapers were thus considered an alternative to low-level push TV. The CSC Standard Program in 1976 consisted of tagged TV advertising at 40 GRP per week for 6 weeks in 54 markets, plus the newspaper test discussed above. Advertising in most markets started April 12th in two-week flights with two one-week breaks, for a total of 6 weeks lasting through June 5th. Since the standard program ran in all USDA markets, it was impossible to judge its effects in comparison with markets not receiving the program. The study concluded that, "Due to the rise in crop value experienced in 1976, it was logical to assume that the program had been instrumental in increasing demand." Production increased by 5% while farm value rose 13%.

Table 18
Strawberry Features in Newspaper Test Markets: 4/4–6/5/1976

	Avg. Sq. Cm./Wk	# Ads / Week	Avg. Price Per Test Market
Newspapers Alone			
Boston/Providence	147	2.7	50¢
Atlanta	79	2.7	53¢
Minneapolis	90	2.6	49¢
Salt Lake City	86	1.1	36¢
	101	2.28	47¢
Newspapers & TV			
Buffalo	120	1.7	47¢
Nashville	66	2.6	47¢
Milwaukee	229	3.6	54¢
Denver	89	1.4	45¢
	126	2.33	48¢
TV (Control)			
Pittsburgh	59	1.8	53¢
Birmingham	52	1.3	49¢
Kansas City	139	3.7	39¢
Portland	64	1.6	39¢
	79	2.1	45¢

In 1977, the Board was influenced by the increased price, farm value, and unloads in heavy vs. light TV markets, and budgeted for 14 additional heavy weight pull TV markets, which accounted for almost one third of California volume. Six weeks of TV was scheduled over an eight-week period during the peak season or Stage III. low weight push advertising was still used to encourage retail cooperation in 44 additional markets, with magazine and other publicity efforts, and tie-ins, which will be covered in the analysis of push activities, and were heavily relied upon from 1990 to 2004. This was the largest and most extensive advertising/promotion budget ever undertaken, with $600,000 on consumer advertising and $200,000 for other types of merchandising activities. (The Ad incentive programs did not begin until 1990). The results of the 1977 program are illustrated in Table 18 and represent a 21% increase in volume and a 3% rise in price. Additionally, for the February-July period, heavy TV markets showed a 7% volume advantage over the light-weight TV markets. This fact appeared to indicate that demand had increased, and provided the impetus for further TV expansion. The Board always was aware that strawberries are the first fresh fruit on the market and that there is always considerable feature activity even if no advertising and promotion took place. This was the reason the gradual approach was taken to advertising/promotion expansion and why the best-known marketing research was utilized. We demanded that barometers be used to measure the effectiveness of our programs and emphasized the importance of *price, and net farm income, as guiding barometers for a successful program.*

In 1978, the first million-dollar program was launched. CSC entered into the arena of network TV and heavy weight TV programming at 100 GRP, the weight level tested and proved effective in 1976, was budgeted to virtually cover the entire United States. For 5 weeks in April and May, the peak production weeks, the country would be exposed to a major advertising program, comparable to non-produce type consumer advertising. Additionally, for 3 more weeks, 15 extra markets would enable TV to provide continuous support from mid-April to late June. The combined network and spot schedule would reach 75% of women an average of 6.3 times over 8 weeks. Three top Canadian markets, Toronto, Montreal, and Winnipeg, were scheduled for 6 weeks of TV, at 100 gross rating points during the peak season.

Unfortunately, due to heavy rains, production per acre declined from 22.4 to 19.4 tons per acre, and farm value dropped 6¢ per lb. Nevertheless, the FOB price increased 10¢ per 10 lb. crate, and the board continued Network TV through 1979–1980, as prices and farm value rose. However, no further testing was performed and no other barometers were applied to program effectiveness. In 1981–82 acreage stabilized at around 11,000 acres and prices continued to inch up and farm value followed acreage and per acre yields. Meanwhile, TV had become more and more expensive as had the other consumer, retail, and food service programs. CSC revenues, which depended on volume and the mandatory assessment rate, were inadequate for the magnitude and breadth of the programs and were increased to the maximum 5¢ per fresh and freezer crate to cover the industry's projected growth and to enable the CSC to keep up with inflationary costs. Although during the 1972–82 decade, *price to the grower had increased over 80%, real prices (i.e., prices adjusted for inflation) had remained virtually constant* (See Tables 25 & 27, USDA study). The industry investment in pomology and horticulture enabled production per acre to increase and provide the increase in net farm income. Increased demand was mandatory with projected acreage increases, if prices were to rise. In addition to TV, other push and pull programs were expanded and the board initiated a *Strategic Marketing Plan* to evaluate the effects of the CSC's large, multifaceted advertising/promotion expenditure in 1982.

The CSC had been involved with push marketing techniques since 1968, when the CSC increased the mandatory assessment rate from ½¢ to ¾¢ per tray in order to launch additional research expenditures and to begin promotional push activities. Push marketing is the traditional commodity board approach and is designed to encourage retailers or food service users to feature and display a given commodity because their sales and profits would increase. Some of the techniques are:

1. <u>Token Consumer Advertising</u>: This is usually low weight TV, radio, or newspaper media advertising; ads are tagged with the names of the chains that agree to feature and/or display the commodity.

2. <u>Point of Purchase (POP) Material</u>: This technique offers the trade attractive materials for in-store use in building displays and attracting consumer attention within the store.

3. <u>Display Contests and Incentives</u>: This is a direct approach to the produce manager where prizes are given for building a display. To be successful the prizes must be interesting to the retailer and provides the retailer with a chance of winning. The prize list and contest rules must be communicated throughout the country.

4. <u>Trade Advertising</u>: This is also a direct approach where advertising in trade journals, such as the Packer, informs the trade of the promotion program and obtains their cooperation.

5. <u>Merchandising Field Force</u>: The CSC and other Boards provide this service to implement programs adopted by the industries such as the *category management* program.

6. <u>Advertising Promotion Incentives</u>: This technique utilizes direct payments to retailers for advertising and will be treated separately as the most significant CSC program affecting the demand-shift variable, CSCPROMO. This program was adopted by the CSC in 1990 and continued annually until its termination in 2003.

The Strategic Marketing Plan mentioned above included the pull techniques, including consumer advertising, public relations, consumer sweepstakes or contests, and couponing. Before evaluating the studies results, a brief outline of the other push and pull techniques used between 1968 and 1986 will further illustrate the use of *other* barometers to measure the effectiveness of programs, *beside the price and farm income barometers.*

The mailing of simple point-of-sale kits was begun in 1968 and was considered the most ambitious promotion program in the history of CSC advertising. Price cards featuring full-color reproductions of strawberry shortcake were mailed to retailers along with *bin-strips* from the prior year. Two consumer programs were launched with color photographs of strawberries to major market newspapers for printing on food sections. Materials were also supplied to leading ladies' magazines. Radio and TV homemaking programs were sent a steady stream of easy to demonstrate ideas on serving strawberries. By 1970, retail display contests with large prize offerings were expanded including a vacation trip to Hawaii for the top display of strawberries. In the 1968–1970 years the average entrant in the contest had reported a near doubling of sales. There were another 118 prizes for displays including TV stereo and radios. This was also the year of the joint Pillsbury-Kraft promotion, combining the efforts of 3 participants in two supermarket departments, to benefit both the produce and dairy case display departments. Pillsbury Hungry Jack biscuit dough, Kraft Whipped Topping, and strawberries were combined in a shortcake promotion effort. Additionally, Kraft's 2,500 field men across the country were working directly with retail produce and dairy case department managers in building attractive displays. Full-page ads combining recipes and pictures of berries, biscuits, and topping appeared in all leading ladies' magazines featuring strawberry shortcake. When TV was launched in 1974, retail tag announcements were encouraged for individual chains on both TV and radio at the end of each 30-second commercial. For the first time, a 5-man field staff operating under the CSC began calling on retail produce merchandisers throughout the United States and Canada to assist them in preparing for the new TV ads, other expanded display contests and prizes, and point of sale materials.

The board was impressed with the reported sales gains by stores throughout the country. Examples were a Safeway store in Denver, where sales increased 200%, and an IGA Foodliner store in Perry, Missouri, where sales in one day exceeded sales of a normal 2-week period. Many companion items were displayed and sold with fresh California strawberries in contest displays. The A&P stores throughout Maine and New York had increased sales of heavy cream, Cool Whip, Bisquick, angel food cake, pound cake, and ice cream.

The 20th annual *Frozen Food Age* magazine, one of those periodicals used by the CSC for public relations advertising of frozen strawberries, reported that the leading seller in the frozen fruit section was a 10 oz. package of sliced strawberries. It was also reported that King Super, in Denver, increased its store allocation of frozen strawberries to 35% from 15%.

Many other tie-in programs began in 1976 with an offer of a 50¢ refund on strawberries with a proof of purchase receipt from strawberries and a box bottom from Bisquick. Also, Certo and Sure-Jell fruit pectins spotlighted strawberries during a major TV and print media schedule aimed at new users. Supported by extensive advertising and in-store display material in the three Pacific Coast states, three California commodity groups cooperated in a fall promotion program which gave consumers appetizing new uses for strawberries, sour cream, and cling peaches. The CSC, Cling Peach Board, and the Milk Advisory Board combined efforts to reach consumers through media and point-of-purchase materials providing beautiful recipes. The idea was to continue and broaden the heavy CSC program, begun earlier in the season, into the summer and fall months when there was an abundance of substitute fruits.

In 1976, supermarket check stands featured copies of *Family Circle* in August, and *Woman's Day* in September, along with *Sunset* magazine in order to reach the reader whose demographics parallel the profile of the heavy strawberry user. The ads stressed the nutritional value of California strawberries, stating that as an excellent source of Vitamin C, one cup fruit supplies about 150% of the US Recommended Daily Allowance for the average adult.

Couponing was another push technique previously used, and expanded in the 1976 campaign The CSC joint Strawberry Shortcake Festival promotion with Hunt-Wesson's Reddi-Whip and Stouffer's frozen pound cake. The promotion involved the distribution of 7 million free coupons good for 50¢ off the supermarket purchase price of strawberries. In addition, the promotion supported retail sales of all three products with heavy consumer advertising on TV, in newspapers, and in high demographic editions of *Ladies Home Journal, McCall's,* and *Better Homes and Gardens.* A strong food page publicity program helped to call attention to this buying opportunity. The theme for all of this effort was, "Its Either Now or Next Year."

The above narrative is a brief synopsis of the beginning of the advertising/promotion media journey, Phase 2, mentioned in the earlier section on Marketing. In 1982, Dr. Dale D. Achabal of the University of Santa Clara prepared the Strategic Marketing Plan to provide basic guidelines in evaluating marketing alternatives for the board. The CSC board desired research to indicate the relative effectiveness of the push and pull programs over a period of time with barometers for measurement. The Plan addressed itself to the promotion and distribution aspects of the marketing mix. Production was considered to be the province of the grower and processors, although the CSC does participate in research and development. *Pricing* is solely a function of the interaction among shipper and processor sales organizations and market conditions and is only, as Achabal states, "Indirectly affected by CSC activities." Therefore, Achabal's Strategic Plan recommended that consumption for fresh and frozen strawberries be augmented by protecting the large user base for fresh strawberries (estimated at 86%) and increase the consumption among light, medium, and heavy users. More specifically, the goals and objectives of the CSC should be:

1. Maintain or expand consumer user base at the 1982 level. This will entail capturing at least two-thirds of the increase in households between 1982 and 1987.

2. Increase consumption of fresh strawberries by low and medium users, as defined in the 1982 Consumer Marketing Research.

3. Increase percentage of current light and medium users who consume strawberries alone, as a snack, or in simple preparations, as measured by the 1982 Market facts Panel.

4. Increase the percentage of supermarkets using CSC points of sale material from 15 to 25.

5. Reduce the incidence of low or out of stock conditions during peak shopping hours from 40% to a maximum of 25%.

6. Halt the decline in retail strawberry ad features and increase them.

The barometers for measuring fresh program effectiveness are the following:

1. Monitor Simmons (SRMC) consumer user base for strawberries on an annual basis.

2. Conduct periodic consumer research studies using Market Facts Panel to monitor household consumption, usage, product recall and attitudes and awareness toward fresh strawberries.

3. Conduct CSC field staff store audits to monitor POP material usage and out of stock condition in a national sample of supermarkets.

4. Analyze retail food page advertising using Majors Grocery Ad Books to determine trends in fresh strawberry ad features.

5. Monitor fresh strawberry sales as a percentage of gross produce department sales (1982 was 2.7%).

If we assume that demand is mostly a function of price, the other demand-shift variables take center stage and the barometers mentioned above are a method of measuring the effectiveness of CSC fresh strawberry programs directly on strawberry sales, or as proxies for sales.

Frozen Retail Usage

During this 1972–82 period, the overall consumer pack of frozen strawberries represented 18% of frozen production and approximately 5% of total tonnage produced in California. Retail merchandizing of frozen berries was characterized by limited display space and visibility. The advertising for frozen strawberries by retail stores was at a relatively low level. However, CSC testing in 1981 indicated that significant increase in retail store advertising of approximately 42% could be achieved through cooperative advertising efforts. The research study indicated that a relatively small user base exists for frozen berries. Only 55% percent of households in the survey reported purchasing strawberries during the past 12 months, and usage levels for low, medium, and heavy frozen users were identified at significantly lower levels than fresh. The survey did not indicate consumer dislike for berries, but simply a lack of "top of mind awareness" of frozen usage as a fresh product alternative. SAMI (Science and Math Initiatives database) data showed an overall decline in the frozen fruit category, which was considered a possible limiting factor for growth of strawberry sales in this segment of the consumer market. A major opportunity for increased retail sales of frozen berries was to generate trial usage and broaden the consumer user base. A specific goal was to expand the user base from 55% in 1982 to 60% by 1987. Another *barometer of increased demand* could be the increased usage level of current users through increasing product purchase frequency from an average of 1.9 times per household monthly in 1982 to 2.4 times by 1987. The CSC strategy was to focus on specific usage ideas in consumer media through advertising and public relations. The focus of promotional effort was directed toward late fall and winter months to stimulate off-season sales in order to reduce processor inventory carrying costs.

The suggested barometers for measuring the possible increase in demand would include monitoring the consumer user base and frequency of purchase as part of periodic Market Facts Panel surveys, utilizing *Frozen Food Age* and SAMI data to monitor shifts in consumer retail sales, and measuring awareness of consumer advertising consistent with expenditure levels.

Fresh Food Service

The CSC food service programs from 1972–82 involved 20%-25% of the berries consumed. As much as 50% of the volume is used in the food service industry during the later part of the season. The goals and objectives of the CSC food service programs were to: maintain market penetration at 62% and increase utilization from 1,500 to 2,000 pints/units of strawberries in the white tablecloth segment, maintain market penetration (63%) and utilization (7,400 pints/units) of berries in the hotel/motel segment, increase penetration in the family service segment from 38% to 42%, and increase penetration and usage in the contract feeder segment. The strategies for accomplishing these objectives were to orient advertising and publicity toward the white tablecloth and hotel/motel segment as well more comprehensive support systems for fresh produce distributor organizations. The initial benchmark for measuring success in the food service segment should be provided by Technomics International research firm.

Frozen Food Service Usage

The frozen strawberry food service market segment has traditionally been a substantial portion of the frozen strawberry pack. In 1982 it was estimated that 21% of the frozen pack went into food service containers. The retail pack, which is predominantly *private label*, is held in inventory much longer than the food service pack and can be used only at the retailer's demand, while the food service pack provides greater flexibility in marketing. The major opportunity for growth was in penetrating the domestic food service segment; namely increasing the usage of Coffee Shop/Family Service from 29% to 35% and also increasing usage levels in the Institutional and Contract Feeder segment by 1987. The strategy to achieve penetration was CSC concentration on multi-unit food service operators and to continue developing cooperation between distributors' sales organizations and key institutional users.

The initial benchmark for measuring CSC success in the food service segment was to be the 1978 Technomics research study. In the years from 1978 to 1982, the food service industry experienced its most significant growth and a new survey with revised benchmarks was necessary but never completed.

Consumer Analysis: 1982–1998

Comparative data for all of the barometers is not available, but we have examples of relative changes in some barometers for 1982–1998 that measure the total effects of the push and pull demand shift variables outlined above, not the specific effects of each program. As we mentioned earlier, the period from 1976 to 1990 represented the *middle* experimental phase of CSC demand shift programs and this study provides proxies for sales, such as usage, consumer attitudes, distribution of fresh and frozen strawberries, shopping behavior, and purchasing patterns by container size. The primary purpose of the Achabal studies, 1982,

1986, 1992, and 1998, was to better *understand buyer behavior* as it relates to fresh strawberry usage patterns, including purchasing behavior and perceptions of the product, and how these are changing over time. The following Table 19 illustrates fresh usage levels over time and is a barometer for the *measurement of effectiveness* for total CSC advertising/promotion programs:

Table 19
Fresh Usage Levels: 1982–1998

	1982 %	1986 %	1992 %	1998 %
Non-User	14.1	17.1	5.8	6.3
Light User	28.2	23.7	31.1	30.8
Medium User	27.1	26.5	33.2	31.4
Heavy User	30.6	32.7	29.8	31.4
User Base	86	83	94	94

These numbers represent a continuing high user base, but more importantly, non-users have become light and medium users since 1986. A commodity board might conclude that the mixture of push and pull programs has caused the user base to expand resulting in greater proxies for sales. It is *impossible to assess* whether the CSC programs or the increased supply was the reason for the increased user base. The study did not address this cause and effect relationship. It is important to note that it must be assumed that variables such as usage, product recall and awareness, and consumer's attitudes are highly correlated with actual sales. This assumption may or may not be valid depending on circumstances (Jang-Ling Lee). However this data was one positive factor that caused the CSC to continue a large advertising/promotion program. Other barometers in the study that provided positive and negative indications to the CSC of their advertising/promotion programs effectiveness were:

1. The user base for frozen strawberries was 5.3% in 1992 & 1998, 7.6% in 1986 and 13% in 1982, a continuing decrease, which raises questions about CSC freezer program effectiveness.
2. In 1998 and in the three previous surveys, *solo* and *ingredient* (terms used in the Achabal study) uses dominated household consumption

because these segments were the heaviest users. An advertising program begun in 1983 provided the consumers and retailers with the catchy slogan, "Great straight," which *may* have been a reason for the solo success.

3. In the survey conducted from 1982–1998, the characterization of a person who eats and enjoys strawberries has remained stable. The description of a person who enjoys fresh strawberries continues to be very positive indicating that strawberries continue to have a positive symbolic overtone. The facts indicated by the long term survey that people like fresh strawberries, are health conscious and enjoy eating healthy food, may have been due to CSC public relations and improved quality.

4. In 1982 and until 1998, there had been a considerable decline in the survey respondents who disagreed with the statement "best strawberries come from California," specifically 29%—1982, 27%—1986, 19%—1992, and 20%—1998. Quality, large increases in California supply, and the reduction of acreage in other growing districts could have been the reason for this decline, but the CSC public relations and retailer's unwillingness to stock local berries are other reasons for positive consideration of push programs by the CSC.

5. Public relations, newspaper and magazine advertising and food articles, and retail promotion of nutrition have probably influenced 76% of consumers of whom the statements, "low in calories and rich in Vitamin C" describe fresh strawberries "very well." This should mean that the CSC nutrition message and programs designed to influence consumer demand are somewhat effective.

6. Other positive attitudes among consumers over this time period are indicated by the decrease in consumers who believe strawberries are more expensive than other fruits, are a luxury fruit, spoil more quickly than other fruit, and are difficult to find except in the spring. Again, the CSC advertising/promotion push and pull programs have been designed to influence these *measures of effectiveness* and ultimately the demand for strawberries.

7. After the price variable discussed earlier, the most important variable influencing the purchase (demand) of fresh strawberries is *perceived flavor*. 95% of consumers surveyed believed that *taste* was the most important consideration in a purchase, followed by color and appearance. The two next most important attributes influencing the purchase

of fresh strawberries were price and retail displays, followed by uniform shape and packaging. The CSC has influenced *displays*, and *packaging* directly through research and merchandisers and the relative success of each are barometers of their programs. Flavor, color, and appearance are influenced directly through CSC-UCD research in pomology and horticulture. It is in this important area that the Achabal survey does not measure the consumer attitude during 1982–1998. My research and experience indicates that the non-proprietary grower/shippers are competitively superior in taste, color, and appearance, compared to those entities growing university varieties. The research does indicate that consumers shop a particular market because fresh strawberries were of better quality. In 1998, 54% of all strawberry users shopped a partic- ular market for quality, representing a slight increase from 1982, 1986, and 1992, where the corresponding figures were 50%, 46%, and 42% respectively.

8. Although retail newspaper ads are discussed in a later section on the variable, PCPROMO, the following consumer survey indicates that an *advertised special* on fresh strawberries causes the consumer to shop in those retail markets involved with the specials.

Ad Purchase Pattern:

	1982	1986	1992	1998
Often Purchases	29.5%	30.1%	45.8%	44.7%

This consumer survey provides consumer behavioral information that indicates their purchase pattern follows retailers that advertise specials on strawberries. Boards can use this information to encourage retail advertising for competitive purposes with the knowledge that advertising specials increase demand. Ads are one of the best barometers for indi- cating increased demand.

9. The necessity for retailers to display strawberries and board merchandis- ers to *track and quantify displays* are essential barometers of a program's success, because 60% of the survey respondents buy strawberries when they see them in the store. CSC programs designed to increase displays have been successful, and the number and size of displays impacts demand considerably. Tracking displays, product positioning, and out of stock conditions are barometers for measuring both the CSC and private sector program success affecting the demand shift variable, CSCPROMO.

10. The consumer attitude on pesticide uses on fresh strawberries is another barometer of attitudes effect on demand. The survey only covers 1998, but indicates that 55% of the respondents are concerned about the use of pesticides, while 62% were *confidant fresh ones are safe*. Additionally, consumers felt that the benefits of eating fresh strawberries outweigh any currently perceived health threats. Tracking attitudes of this *safety issue* over time, as well as other related health questions, are important barometers of possible shifts in demand.

The Achabal studies became the basis for programs implemented during 1982 through 1990. These were the media years, but also years of growth, experimentation and study. Grower returns and prices were not keeping up with inflationary costs and once again the increasing cost of TV caused the board to reassess its marketing options. Proxies for sales were the prime barometers for *measurements of effectiveness.*

The continued acreage increase from 11,000 in 1980 to 20,000 in 1990 was the basic reason for implementing the *Achabal Strategic Plan*, which utilized all of the push and pull techniques that were implemented in the 1970s with some new additions and relative budget changes, emphasizing different programs. The specific marketing goal in 1982 was to market an additional 5 million trays of fresh strawberries without a loss in market value per unit. The projected increases in acreage and productivity from the new varieties caused the CSC to continue stimulation of demand in order to prevent a loss in value to the grower. The previous marketing success for the 1975–81 period, during which fresh shipments increased from 269 million pounds in 1975 to 381 million pounds in 1981 with crop value doubling and FOB prices $2 higher, was unusual in perishable commodity marketing because volume and price per unit increased simultaneously.

The CSC, continued its consumer advertising efforts in 1982–84, however Network radio covered the entire country and spot TV in 19 markets delivered the new *Strawberry Good Ideas* message. The increasing cost of TV required the change in direction, although the greater effect of TV was recognized. The 19 cities were chosen because they were known to have high unloads of strawberries. The main radio campaign ran for 6 weeks during the peak months of April and May, but additional radio campaigns ran in June to support summer volume and in October to support frozen strawberry sales. Meanwhile, trade communications hit new highs with a newly revised strawberry merchandising slide show utilized by over 80 independents and chains to help train their produce personnel to maximize strawberry profits, the emphasis being to reduce out of stock conditions and develop proper displays, which translate into greatly increased total store sales. The CSC field staff was increased to 5 full time employees with responsibilities to encourage the use of point of sale material, provide proper care and handling instructions, distribute the training slide show, and encourage produce departments to merchandise strawberries. Support for frozen berries was expanded with radio ads, trade communications and in-store merchandising. Point of sale material was also distributed to retailers throughout the country. The average quoted FOB price remained about the same and crop value was close to the 1982 record. The CSC continued to use price and farm value as barometers of program success. Other barometers of measurement of effectiveness were the Produce Marketing Association data, which indicated that strawberries now represented 3.69% of total produce department sales compared to 2.81% a year earlier. Another barometer was the number of retail ads, which increased 224% in April, declined 48% in May and in June rebounded with a 48% increase, compared to the previous year. The results of the study by Technomics International consultants showed a dramatic increase in strawberry use in all segments of the food service market between 1978 and 1983. The most significant increases occurred in the coffee shop/family service segment, 89%, and hotel/motel, 51%. The fifth annual CSC promotion contest for food service operators was won by the COCO'S chain for their 7-week promotion, which increased berry use from 72 trays to 2,500 trays during the promotion period.

During 1983–84, a new series of 10-second TV spots were produced with the theme "California Strawberries, they're great straight." This was to capitalize on the change in the consumption pattern discovered in the Achabal study, which indicated a change in strawberry predominant pie and shortcake usage to a simple out of hand or snack fruit. The research showed that heavy strawberry users preferred berries straight, so the advertising theme was to persuade light and moderate users to think of strawberries the way heavy users do. With the change in media strategy the CSC was able to significantly increase *the number of consumer impressions* with only a slight increase in spending. Following the 1984 advertising schedule the CSC commissioned another study to measure the effects of the new theme. Walker Research surveyed consumers, where advertising was heavy and also 6 control markets with no advertising, and found that awareness of TV advertising for strawberries was significantly higher in the test than in the control markets. Of all possible responses to the question, twice as many people in the test markets said they like strawberries because you could eat them plain, generating awareness of a key strategy point. This was another example of a barometer (a measurable behavioral change), a proxy for sales, and a possible change in a demand shift variable, CSCPROMO, from a specific program.

Another CSC survey confirmed that beverages are the number one use for frozen strawberries in foodservice and with this information the CSC published a Beverage Guide for operators and bartenders to assist them with customized promotion for fresh and frozen strawberries. The sixth annual Ripe Success and Juicy Profits promotion contest was recognized as the outstanding promotion for 1984. The Westin St Francis Hotel in San Francisco was chosen because fresh berries were served to their guests from checkin to checkout. This is another *measure of effectiveness* for a CSC program. Merchandising efforts by the CSC 5 field personnel, based on retail strawberry profitability, included a continued retail-training program relying on store surveys showing the importance of front entrance and dry table/end aisle displays, the berry appearance, size of displays and amount of product. Data collected by the CSC stressed that front entrance and end aisle displays increased form 1982 to 1984 and out of stock conditions dropped to less than 5% in the stores surveyed from 13% in 1982. The effect of displays, aisle location, and ad features, *all barometers* for measuring the demand shift variable, CSCPROMO, has shown that customers will buy 76% more fruit than the normal impulse buyer.

In 1985, the CSC increased visibility by including outdoor billboards in several key markets in the US and Canada. The strategy was to post billboards close to major supermarkets and tag them with store logos. Follow up awareness testing showed that the outdoor advertising increased awareness of strawberries. During the peak season, April, May and June, TV covered 22 major markets and retailers were advised to key their advertising/features to coincide with the volume. However, during the peak months returns to the grower were disappointing and the question of program effectiveness, while positive when analyzing other barometers of program effectiveness, were of little importance to grower/shipper members of the CSC who experienced lower returns. Whether returns were higher than they would have been without CSC programs is an unanswerable question and depends upon the evaluation of a majority of the industry paying mandatory fees to the CSC.

All of the other push and pull techniques were part of the advertising/promotion mix for 1986–1988, but the expense of TV and the relatively flat or declining farm value caused the CSC to conduct a comprehensive test of TV effectiveness in 1987 (See Tables 25 through 29). Meanwhile, acreage increased from 14,600 in 1985, to 17,650 in 1988, and was projected at 20,000 acres by 1990. The *perfect storm* of increasing production per acre in addition to rising acreage required bold action; armed with positive TV test results the CSC ventured back into TV advertising, but in fewer markets. The board "chucked" radio and billboard campaigns and focused its entire consumer advertising on TV. Commercials reintroduced to TV after a 2-year hiatus were backed by musical jingles and voice-overs by actor, Burgess Meredith. Targeted for cable TV and spot TV in 8 markets, the commercial centers on the CSC's new campaign, "Strawberries cause spring." The test showed households that buy strawberries would purchase 12% more product after seeing the commercials. The evidence seemed compelling. Although the billboards were popular merchandising tools with retailers, particularly in larger cities, and less costly than TV, *the inability to prove effectiveness* during their 4 year life caused CSC discontinuance. The development of the day neutral Selva variety and a peak production pattern from June through October required the CSC to assess advertising/promotion programs that would help strawberries compete with summer fruits. The new Berry Patch program encouraged retailers to combine all summer berries in a single more prominent display of strawberries, raspberries and blackberries. This display strategy allowed for flexibility in merchandising and advertising depending on availability of each berry, and did not require a rearranging of the department. Store check data showed the average strawberry display alone in June, July, and August was 13.3 square feet, but the average Berry Patch display was 22 square feet. Also positioning in the department and use of POP material was substantially better in the 54% of stores featuring the Berry Patch display. Although TV was given a high priority for pulling by trying to influence the consumer, the push programs continued. The CSC consumer, merchandising, and food service programs continued the 10-year continuity of expenditures at $1 million, $900,000 and $800,000, respectively. Retail training efforts climbed the priority list and the CSC 4 person merchandising staff met with retailers and their store-level staff encouraging colorful displays. An 8-minute video on strawberry merchandising and handling, including bulk and Berry Patch displays, was updated for showing to retailers. Other merchandising activities included the distribution of a merchandising binder containing new materials, newsletters, sales bulletins, ad slicks, research findings, and tie-in opportunities. Perhaps the most important of all was the *expansion of incentive promotions involving sales and display contests*. CSC consumer public relations activities included new press kit covers and letterheads aimed at food editors. Meeting with media continued to encourage features and articles on California strawberries. Consumer magazines continued to be one of the most effective ways of stimulating interest and sales. Because of their beauty and versatility, few products get as much food page and magazine coverage as strawberries and CSC staff provided magazine and newspaper editors and syndicated columnists with recipes and feature story ideas. The media was a major recipient of CSC food safety information including industry's efforts to reduce the use of pesticides. It also reminded everyone of the importance of a diet rich in fresh fruits, including strawberries. All of this information was in a Food Safety Kit provided to retailers in order to help them insure greater consumer confidence. Food service sales reached $215 billion by 1989 and real growth averaged 2% annually for the prior 5 years compared to annual retail growth of 6%. The CSC has been continually involved in developing increased strawberry penetration into the food service and industrial sectors. A 1989 survey of 200 restaurants owners and managers sponsored by the United Fresh Fruit and Vegetable Association found strawberries to be the most profitable produce menu item; 74% of them cited strawberries as the one item that most increased profits. New fresh and frozen ads in leading restaurant publications ran in 1989, along with direct mail, newsletters, and public relations that supported the California Quality theme. Dramatic new recipe cards for fresh and frozen berries were also featured. United Airlines, Coco's Family restaurants and Hilton Hotels were examples of noteworthy promotions. The above historical chronicle and evaluation of CSC programs is designed to indicate examples of proxies for sales. The proxies provided many qualitative barometers of the effectiveness of CSC programs, on the demand shift variable, CSCPROMO, during the media years.

Illustrations on the following pages: Some of the past and present CSC Chairmen, Board Members and others.

*Picture 18. Top: Dr. Royce Bringhurst, center, with Herb Baum and George Murai.
2nd Row: George Murai, Fritz Koontz, Dave Riggs, Ed Kelley, Bill Ito.
3rd Row: Dave Riggs, George Yamamoto, Bill Moncovich, Ken Morena, Red Bryan
Bottom: PJ Mecozzi, Roger Wood, Mark Murai, Richard Amirsehhi.*

Picture 19. Top: A.G. Kawamura, Luis Chavez, Abel Maldonado, Harry Watanabe
2nd Row: Bill Ito, Mack Ramsey, Jim Clark, Cesario Ramirez, Jean Downes
3rd Row: Ed Kelley, Charlie Iwanaga, Sandra Grcich, David Kirk
Bottom: Efraim Contreras, Luis Chavez, Reid Wagstaff, Mark Murai, Neil Nagata

Picture 20. Top: Paul Murai, Doug Circle, A.G. Kawamura, Harry Watanabe, George Chavez
2nd Row: Craig Kotake, Steve Yamamoto, Luis Chavez, Efraim Contreras
3rd Row: Scott Deardorff, Howard Tsukizi, Fritz Koontz, Ed Ortega
Bottom: Peter Navarro, John Dullam, Jack Burke

Picture 21. Top: PJ Mecozzi, Doug Mita, Jose Corona, John Magarro, Steve Garrett
Center: Craig Moriyama, Paul McHaney
Bottom: Garland Reiter, Scott Deardorff, Ron Burke

Picture 22. Top: Sam Gabriel, Louis Ivanovich, Gary Wishnatzki
2nd Row: Ron Uesugi, Paul Tognazzini, Teri Tognazzini, Vince Solbes
3rd Row: Richard Ueymatsu, Sharon Tognazzini
4th Row: Cecil Martinez, Harry Watanabe, Ed Kurtz, Paul Murai
Bottom: Randy Ito, Charlie Iwanaga

Picture 23. Top: Pat Riordan, Bill Moncovich
Bottom: Ira Nathel, Herb Baum

Picture 24. Top: Watsonville grower Gilbert Yerena and Gary Wishnatzki;
Bottom: Alvin Nathel

CSC Incentive Programs

In 1990, a major shift in CSC programs and emphasis began. The major shift within the composition of the demand shift variable, CSCPROMO, was the movement toward the *promotion incentive segment* of the other advertising/promotion programs, excluded from our previous analysis. We chronicled other CSC promotion programs encouraging retailers and food service operators to advertise, feature, and promote strawberries, but did not introduce the analysis of incentive promotion earlier, because it did not begin until 1990. The CSC Board recognized that price, yield per acre, and farm value were relatively constant as acreage increased from 11,000 to 20,000 acres during the 1976–1990 media period. Meanwhile, production costs were increasing at least 20% over the 10-year period, profitability was declining and growers were concerned about the direction and emphasis of CSC activities. The CSC realized the importance of consumers and 1990 research by US Marketing Services indicated that strawberries make a substantial contribution in the retailer's battle for market share and profits. The new generation of strawberry customers typically spends more money on produce and throughout the store than those who don't buy strawberries. To seize this opportunity, retailers needed to attract strawberry customers and merchandise strawberries to keep them satisfied. The inflationary cost of TV, the necessity for a minimum of 150 GPR to "break through the clutter," a limited CSC budget and decreasing grower returns caused the board to utilize the push approach of encouraging the retailer to push strawberries through incentive programs rather than media consumer pull programs. In addition to the expense of TV, the popularity and importance of strawberries to the retailer and consumer persuaded the CSC to simply remind the retailers that strawberries are available, healthy, tasty, and profitable to handle, thereby utilizing the push techniques and discontinuing TV.

The 1990 spring ad incentive program began with CSC sending a program application to all retailers which stated, "The average retail chain runs 3.4 strawberry ads featured in April and May. The CSC will provide a cash incentive for performance above the national average." To qualify for an incentive, the retailer must advertise California strawberries more than the national average of 3 times during April and May and for the fourth and each succeeding feature or sub-feature between April 1 and May 30, 1990. The incentive dollar quantity per ad was to be determined by the retailer's history of advertising, size, and willingness to provide proprietary data on volume. The amount generally ranged from $1,000 to $2,000 per ad over the national average, which changed over the years. The performance guidelines required that a minimum of one ad per week must feature or sub-feature California strawberries, run chain wide, with price-effective dates clearly marked. Additionally, a 1990 Summer Stock incentive program began, recognizing the increased summer and fall volume resulting from day neutral planting. The average retail chain ran 1.5 ads in June and July and the CSC provided an incentive for performance above the national average with all of the rules governing the spring program applying. The CSC incentive program provided $1,000 to $2,000 per ad over each retailer's ad base, and a small percentage of the $10,000 to $15,000 total retailer ad expense. The CSC was not buying an ad for the retailer, simply providing a possible incentive to the retailer to advertise strawberries, which was known to be a food item that receives much advertising and public relations. Retailers were informed that *300 to 500 million impressions* are a substantial strawberry presence in advertising and the result of strong CSC public relations, along with 60 magazine covers, 500 magazine stores and countless newspaper articles. The CSC merchandisers also provided retailers with support information on displays, point of sale material, tie-ins, and data on the loss of sales and profits with out of stock conditions, especially during peak shopping hours. As we indicated earlier the Achabal study shows that consumers are increasingly responding to strawberries as an advertised special, reflecting the consumer drawing power they offer supermarkets. The study provides behavioral barometers (proxies) for measuring the effects of retail ads, along with the important number of ads in Table 8. The increase in the number of ads from 1991 to 2001 are indicative of the increase in volume and relative importance of strawberries, the cumulative effect of media, merchandising (displays, tie-ins, product information), and public relations. Retailers are influenced by competitive practices, including the frequency of ads, and level of pricing, both promotional and regular, and the CSC's category management programs, although all pricing, volume, and ad offers, are purely in the domain of the strawberry shipper and will be discussed later in this chapter in the PSPROMO section. The CSC category management programs include all of the push and pull programs mentioned earlier, but with new and greater prominence for incentive promotion at approximately $500,000 to $650,000 of the total marketing budget of $2.5 million to $3 million. Since a retail ad with a promotion price will increase demand by at least 2–3 times, the CSC goal of increasing ads was appropriate at the time it began in 1990, and an important function for CSC merchandising staff. The ad data supplied by Leemis Market Research, was the data base for retailer information on number of ads, type of ads, color, promoted and regular price, size and prominence, newspaper, in-store, home delivered, family ad, and the dominance of strawberries within the retailer's individual advertising format. This database provided the CSC and shipper/members with a retailer's history over time, enabling the CSC formation of tailored merchandising plans, the ability to measure retailer-CSC cooperation, and information for shipper ad promotions. The CSC incentive programs continued through 2004 and were then discontinued. Since the retail decision to advertise is complex and determined by offers from all produce suppliers as well as other suppliers, the ability to influence a retail ad decision with any monetary amount, let alone the CSC minor contribution, seems relatively insignificant and its discontinuance was a decision that should have been made much earlier. Additionally, the promotion function is and should be between the shipper and retailer.

The CSC continued support of retail, foodservice and industrial promotional efforts with an array of services, including training, POP material, media coverage, market analysis, and a professional team of merchandisers. One of every four meals was consumed away from home representing 45% of total food spending. Foodservice continued to increase due to dramatic changes in lifestyle including more working women, overall need for convenience, and two-income households. These trends caused the CSC to focus on marketing efforts directed toward establishments utilizing strawberries as part of breakfast, lunch, and dinner. Increasing numbers of restaurants were carrying more fresh produce than ever and the CSC could assist and inform them of the importance of strawberries as a decorative, healthy, and economical compliment to food presentation. The CSC provided recipe cards, foodservice publications and quarterly newsletters to keep strawberries *top of mind* with industry leaders. A new Japanese export program was also begun with the help of TEA, Targeted Export Assistance Program, funds provided by the USDA. Japan was the second largest export market for fresh strawberries and *the* largest for frozen strawberries. Data showed that strawberry exports from the US increased 12.2 million pounds in 1990. An important target in Japan was the confectionery industry, which uses 96% of fresh strawberries imported from California. The CSC sponsored a cooking contest with over 100 pastry and hotel chefs entering the competition. California strawberry exports to Japan played an important role in the late season strawberry market. Because of cultural preferences for Japanese strawberries and transportation costs, virtually no California strawberries were accepted into the Japanese market until after June 15. Therefore, all of the California strawberries exported to Japan are marketed during the last half of the California season. During this time, competition from tree fruits, grapes, and melons severely cuts domestic demand for California strawberries, emphasizing the importance of the Japanese export market. Maintaining and expanding demand for exports to Japan during the late season was critical to the health of the California industry. Also in 1992 the CSC launched a new and expanded retail training program with 50 seminars attended by over 2,000 produce personnel, emphasizing the importance of careful handling and proper merchandise to increase sales. Incentives continued to supplement this activity, as tie-in promotions with strawberry-use products, such as whipped toppings, shortcakes and pie shells were introduced, and Kellogg's Corn Flakes became a tie-in partner with California strawberries featured on the side panel of 15 million cereal boxes nationwide. As I mentioned earlier, research by Market Facts in 1992 indicated the powerful consumer drawing power of strawberries and, compared to 1982 and 1986, showed an increase in usage: 94% of the survey respondents consumed strawberries, 75% made a special trip to the store just to buy strawberries, and 60% said they switched stores to shop where strawberries were an advertised special. Strawberries appeared on 21 magazine covers such as *Family Circle*, *Woman's Day*, *Redbook*, *Gourmet*, and *Bon Appetite*. During the year in the food service sector, CSC encouraged Denny's, Coco's, Bennigan's, and El Torito to develop strawberry breakfasts, specialty drinks, and dessert promotions. In the industrial sector, the CSC helped develop data for accurate nutrition labeling of frozen strawberry products, in order to meet the new FDA labeling requirements. Despite all of this marketing activity and a 3,000 acre increase from 1991 to 1992, production per acre declined from 26 tons to 21 tons, volume slipped, peak season prices were lower, and the number of ads dropped. However, some of the positive proxies for sales barometers for increased demand were increases in exports, breakfast food and topping tie-ins, specialty drinks using strawberries, and the potential long term effects of new nutritional labeling of frozen strawberry products. Consumer surveys indicated an increase in strawberry usage from 1982 to 1986. They also provided the CSC and shippers with data that indicated the retail importance of advertising strawberries, since 75% of consumers made a special trip to the store just to buy berries and 60% said they switched stores to shop where strawberries were an advertised special. These potential behavioral changes could be the result of long-term public relations messages, which discuss the nutritional benefits, as well as the uses of this colorful, tasty, solo, and multi-purpose fruit. The retailers may have been encouraged to increase their strawberry advertising because of this CSC activity.

1993 and 1994 were years of continued CSC push and pull programs, without media, directed at the consumer. Two major goals were to provide retailers with promotional incentives and merchandising ideas for displays, tie-ins, and training for produce personnel to improve berry quality at the store level. Magazine covers featuring strawberries, recipe features in major newspapers, and NBC newsman, Willard Scott, announcing on national TV that "Spring is here because strawberries are in season," were all part of CSC public relations, which generated over 400 million consumer impressions, a figure unmatched by many expensive ad campaigns. The CSC food service program continued to expand usage of fresh and frozen strawberries by providing support service and promotional materials to commercial operations or major restaurant chains, non-commercial operations such as school cafeterias, and food distributors. The CSC also expanded its export-marketing program in Japan through one on one contact with key retailers and distributors, seminars and taste tests, and distribution of literature and POP materials. A measure of the CSC export programs success was the increased share of Japanese imports from 30% in 1982 to 54% in 1992. The popularity of strawberries as an ingredient in ice cream, juices, yogurts, and jellies, was promoted by the CSC through trade publicity, convention attendance and technical bulletins, emphasizing to industrial users that California is the most consistent supplier of high quality frozen strawberries.

Measuring the effectiveness of CSC incentive programs is difficult because the amount of money expended by the CSC is relatively small compared to the total expense of a retail ad, $1,500 to $15,000, which is only one of many factors determining a retailer's advertising policies. We do know that an ad will increase demand by 2 to 3 times. We also know that real FOB prices declined during the 1988–1994 period and recovered from 1995–1998 (See Table 10). Production per acre increased from an average 25 tons per acre to 25–27 tons per acre in the same 10 year period while farm value gradually increased because of rising acreage, production per acre, and higher normative prices (See Table 9). The perfect market storm, caused by predicted increased acreage did not materialize because of increased retail advertising (Table 8), to some small extent CSC incentive ad payments, and the other CSC push and pull programs that indirectly influenced consumer, retail, and foodservice and industrial user behavior. Behavioral barometers, proxies for sales (Achabal), actual ads, and the marketing examples of successful promotions were the tangible assets or measurements of effectiveness for assessing CSC programs thus far. The number of consumer impressions, created by CSC public relations activities, was a newly introduced barometer, proxy for sales, but seems impossible to relate directly to demand, farm income and prices. It may have had a positive effect on demand, but. commodity boards of directors will have to assess its importance, as well as other proxies for sales.

The CSC began a new and important function in 1994. The changes in the government regulatory environment involving soil fumigation, the disposal of solid waste, workers' issues, and food safety, all required CSC's attention. The phaseout of methyl bromide could have a devastating impact, not only on California strawberry production but also on the world's food supply. A new research director was appointed and the CSC became involved in the Crop

Protection Coalition, working to ensure the interests of commodities affected by methyl bromide phaseout in Washington DC. This is discussed later in the Supply Shift Variables section of this chapter. The disposal of solid waste was becoming a critical environmental issue, particularly in the eastern US, making the discarding of odd-sized and wooden pallets increasingly difficult. A palletization working group was formed to explore industry conversion to a new pallet that would be interchangeable for all commodities by having identical dimensions, which would increase reuse possibilities and provide less waste. The CSC Regulatory Enforcement Working Group was also established to help the industry understand the workers protection standards and to promote uniform enforcement of standards to ensure compliance from all segments of the industry. The Group coordinated the bilingual TIPP (Targeted Industries Partnership Program) seminars to help growers understand their obligations and avoid expensive fines. These CSC involvements were directed toward educating the industry and strawberry user; that effort has been undertaken to improve worker and environmental conditions and enhance the consumer perception of the strawberry industry as progressive and interested in its health and welfare. Hopefully, this proxy for sales would have an effect on strawberry usage and impact demand.

In 1995 the CSC broke new ground in produce marketing by launching an infomercial to raise consumer demand, provide new usage ideas, and communicate about modern farming practices. The 30 minute commercial ran nationally in 20 markets and generated an overwhelming 15,000 responses from consumers who were interested in recipes and nutrition. In 1990, the average produce section carried about 250 items and by 1995 over 450 commodities were competing for display space. The infomercial was designed to help strawberries maintain their primary display position and despite competition from the 450 other products, ad frequency still increased by 7% and strawberries maintained their primary display position. An infomercial survey was conducted with 94 female adults in the seven market areas to assess the immediate communicative effects of the new program. They were recruited 14 days before the infomercial viewing and several questions about usage were posed; after viewing the infomercial, their reactions were recorded. Reactions were again measured 24 hours after the viewing. The researchers learned what information was recalled and retained, what content *might* cause viewers to increase usage, and what shifts in attitudes were precipitated by the infomercial. The study revealed that recipe, versatility, and nutrition information might increase usage because top of mind recall was expressed in unusually minute detail. Strawberry versatility, new and different usage, and recipes were still recalled 24 hours after the viewing. The researchers concluded the infomercial effectively communicated several messages that *should* have a favorable impact on consumer attitudes toward strawberry usage. *Fresh Trends* reinforces the effectiveness of marketing efforts by retailers who provide recipes and new usage tips, as 40% of consumers have been influenced to buy specific fresh produce items, as a result. Since 60% of all strawberry sales are on impulse, CSC encourages retailers to display front and center with lots of POP, recipe cards, desert cakes, whipped cream, chocolate and coupons. The CSC merchandisers continued the incentive programs, and public relations to influence newspapers, magazines, and other media to increase consumer impressions to over 450 million. The CSC export marketing activities included radio advertising, personal contact with retail and food service operators in Japan, development of country, specific literature and POP material, and the introduction of frozen strawberry beverages for foodservice operators. Meanwhile, acreage increased slightly, production per acre was down, as was farm value. The April, May and June retail ad data increased slightly with the same number of retail chains, similar to the July, August, and September period. This was probably due to reduced volume.

In 1996 the CSC began emphasizing nutrition and the well documented fact that strawberries were high in Vitamin C and also contain more folic acid than any other fruit. According to the USDA, strawberries have a better total of antioxidants than any other fruit and only 50 calories per serving. The CSC linked the infomercial with other consumer oriented massages with a new web page and in the first 4 months received over 200,000 hits. Foodservice research was continued with foodservice operators to measure usage and penetration in various segments of the restaurant industry to ensure that CSC programs were effective and useful. 52% of the US food dollars are spent outside the home and this market understood that strawberries are their most profitable fruit. The industrial ingredient user base has been expanding and it was estimated that 128 million pounds of frozen strawberries are ultimately used as an ingredient in some food product. That represents 12% of total strawberry production. The CSC maintenance of its premier position in this category was a continuing goal. In.1996 export sales accounted for 10 million trays of fresh strawberries and 52 million pounds of frozen strawberries. The USDA Market Access Program supplemented CSC export programs and helped expand market share for frozen strawberries in spite of increased competition from other countries, especially China. All other programs including advertising, development of country-specific ads and literature, and contests were continued. The second annual All Japan Confectionery Association Recipe Contest was actively promoted for the confectionery trade. Acreage and production increased, production per acre remained the same, and farm value increased slightly. The average number of ads increased for the peak season and declined for the summer months.

With 65% of fresh strawberry volume moving directly through retail operations, the CSC placed a large part of its 1997 resources in the retail program. The regional merchandising staff, through one on one contact, tailored CSC incentive programs and other programs to each customer; these established relationships are helpful for regular CSC activities, as well as when crises arose, such as the Hepatitis A scare in 1997. The goal was to provide more crop information, ad tracking data, and all other information that will be helpful to the retailer, foodservice operator, and industrial user. The CSC continued to provide supplemental point of sale promotional materials, food safety kits, training seminars and videos. Incentives, trade advertising and publicity continued to keep strawberries top of mind, and emphasized nutrition and health information. After a three year run, 1997 marked the termination of the 30 minute infomercial. It was unique in the produce industry when it was introduced and ran 80 times between April and June in six major markets. No information was available about its demise. During the year strawberries were featured in over 20 magazines. In addition, over 30 TV stations throughout the nation featured strawberries on their newscast on the first day of spring and Mother's Day. Export activities focused on expanding fresh demand in Mexico, Japan, Canada, and the United Kingdom, and increasing the user base in Japan for frozen berries. This marked the first year a promotional program was conducted in the United Kingdom after the British strawberry season.

Crisis management took on new industry importance because demand for strawberries could be severely reduced when food safety issues were raised in the US. In April, 1997 an unwarranted food scare became front page news as Hepatitis A was falsely attributed to California strawberries, although California strawberries were not involved in any way. It was necessary for the CSC to respond aggressively and communicate with the media, consumers, retailers, and foodservice distributors. Additionally, the CSC conducted its own pesticide residue monitoring to supplement governmental findings and supported extensive research on integrated pest management. In the area of methyl bromide soil fumigation, the CSC continued to work with state and local officials to communicate the industry's safety record and safety use standards. This safety information was part of the CSC public relations programs with consumers, retailers, and foodservice distributors This was one of the rare years of acreage decline, from 25,200 in 1996 to 22,500 in 1997. However, the Camarosa variety and excellent weather increased per acre yield by 2.7 tons, causing production to be only slightly lower and farm value and prices to be up considerably. The number of retail chains continued to decline, but the average number of ads continued to increase ever so slightly as volume declined.

In 1998–99 acreage climbed back 2,000 acres each year and the CSC's public relations programs continued to be the main vehicle for communicating information about California strawberries directly to the consumer. Constant retail consolidation made it essential to continue developing partnerships with key accounts based on solid marketing data and consolidate all information into the CSC category management_system. The CSC merchandisers, armed with the results of market research, provided retailers with data on the profitability of various packs, a guide to optimize the retailer profit mix using the most popular, one pound consumer pack and retaining more net profits than other produce items. The basic goal of category management was to grow the strawberry category with a coordinated CSC-retail partnership, which analyzed retail packing assortment, pricing, merchandising, and promotional activities including incentive advertising payments, and provided the retailer with fact, based recommendations which would result in increased sales. Weekly sales data was collected and used to generate quarterly category performance reports. This report measured the average weekly volume and sales dollars per store, and reflected the share of category by pack, type, and average retail price. They also provided previous comparisons on volume by pack type and dollar sales, along with a promotional activity analysis and weekly category performance index. Daily sales data was collected from 30 retail chains representing over 3,000 individual stores. Individual retailer information was combined to create a national composite, which was broken into 5 geographical regions. The Food Marketing Institute now estimates that two thirds of supermarket companies have implemented category management plans for produce.

The category management system provides information to retailers, which will help influence their decision to demand more strawberries. Examples of data are: produce sales represent 11.28 % of total supermarket volume and strawberries represent 6.5% of total produce sales during the peak months, April, May, and June, declining to 3.5% during the summer months. Comparatively, the annual contribution of bananas is 6.2% of produce sales, apples—10.5%, citrus—5.9%, and lettuce—5.5%. The high sales contribution of strawberries, during the peak season, emphasizes the importance of strawberries to overall department and store performance. However, strawberries still contribute 3% to 4.2% of department sales during the summer, making it a strong department performer. This data, if available in a time series, would make an excellent barometer of CSC programs success.

The data which is available to us, however, is presented as the Fresh Strawberry Advertising History (See Table 8), covering a 10 year period from 1991 through 2001. As the retail marketplace continues to consolidate, the total number of strawberry ads has remained steady indicating that the remaining retail base is frequently advertising strawberries. In 1991, the CSC tracked 440 accounts in 51 US markets and by 1999 only 290 accounts remained in those markets, or a total decline of 34%. However, during the spring, the total number of ads declined only slightly, by less than 6%, from 2,181 to 2,051 during the same period. This represents a significant increase in the number of ads run by each retail unit. In 1991 the average retailer ran about 7 ads in the course of the entire season, while in 1999 the average retailer ran over 7 ads in the peak season alone and 11 ads for the entire year. The summer months added the greatest number of incremental strawberry ads, as the average chain increased from 1.2 to 3 ads, supporting the late season. A shrinking retail marketplace, increased competition for department and ad space, and the increasing number of ads indicate the importance of the strawberry category, mostly because of volume increases and fruit desirability. During May 1999, volume peaked at 4.5 million trays with an average of 4 million weekly, and remained heavy through the summer because of optimum growing condition and new varieties. Production per acre hit a record 29.4 tons and as a result retailers had the opportunity to increase strawberry sales and category profits. 1999 was another revolutionary year of strawberry packaging, with the domination of the 1 lb. consumer pack in the strawberry mix. The first revolution was in the early 1960's when plastic pints replaced wooden veneer pints. The 1 lb. plastic consumer pack incorporated barcodes, individual shipper trademark label, and represented 61% of packaging offered in the spring and 84% during the summer. This was a 52% increase from the spring of 1998 and 65% during the summer. Pints continued to play a larger role in the Northeast, comprising 55% of the spring retail assortment. However, as the season shifts to summer, the 1 lb. consumer pack increased to 84% of sales because the competition for space from summer fruits caused the produce managers to shift to the strawberry packages that save labor, time, and display space. *Half trays* were also becoming a package of choice for the so called *big box* stores like Costco and Sam's Club, as the consumer base expanded to larger sales units. According to *Fresh Trends* 1999 report, 17% more strawberries were sold when customers are given a choice in packaging and 50% of heavy strawberry users prefer to buy in larger packages. The CSC's role in the packaging revolution was minimal, limited to communicating trends and advantages of multiple packages to the retailer and strawberry industry. The cooperation between shipper, retailer, and packaging industry was responsible for this packaging revolution and exercised an important effect on the demand shift variable, PSPROMO. However, communication, or the sharing of information between category management partners, is the key to better understanding a category's performance.

While CSC's gathering, evaluating, and reporting of information about strawberry category performance, ad incentives, retail training, and point of sale materials were also important, a new point of sale kit was introduced and data indicated that POP material could increase sales by as much as 33%. The CSC public relations program continued to be the main vehicle for communicating information about California strawberries to consumers and foodservice operators with magazine newspapers and other media releases The CSC sponsored a radio tour with a registered dietician to discuss the health benefits of eating strawberries and worked with the American Dietetic Association and the Academy of Family Physicians to spread the word about strawberries as a healthy food. An industry survey indicated that 94% of food service operators said that strawberries were their most profitable produce item. This information along with other category information was part of the public relations program. The foodservice marketing program was capped off with a chef's recipe development event at the Culinary Institute of America at Greystone in Napa Valley. The event comprised of key independent chefs, chain operators, and chain menu research and developers, was created to understand and better serve restaurant customers. The CSC regained the inclusion of frozen strawberries in the USDA school lunch feeding programs. Even though 1999 volume was similar to 1998, the average price was higher because of a later season, higher March and April prices, along with higher late season prices due to shorter supplies. *Daily supply* remains the major price determinant and the chief reason for price fluctuations. Due to the 12 month season, the average seasonal price consists of 4 major districts peaking at different time periods with separate and distinct price averages; this makes annual price comparisons difficult as well as measuring the effects of total CSC programs on price and farm value, both barometers of CSC effectiveness. The use of proxies for sales becomes the other important barometer for measurement of effectiveness.

Beginning a new century, the year 2000 continued the upward acreage trend with a 2,300 increase while the CSC activities were targeted toward research and promotion through marketing programs aimed at consumers, retailers, foodservice operators, export markets, and industrial users. The CSC sustained its ad incentive program at $650,000 to $700,000 out of a $1.5 million consumer marketing budget and it remained the *work horse* of CSC programs. *The ad tracking data is impressive but is this success a result of CSC incentive and other programs, or simply increased volume and proprietary sales and promotional activity?* This is discussed later in this chapter in the PSPROMO-CP section. The fact remains that in 1991 the CSC tracked 440 retail chains, which ran a total 3,234 strawberry ads between March and September; however, in 2000, the number of retail chains had declined to 278 with the number of ads increasing to 4,270. Included in those figures, the April, May, and June ads increased from 1999 with 1,634 ads to the year 2000 with 2,016 ads, while late season ads increased from 3 to 4.5. We know from this data that ads increase demand from 2-3 times. Therefore, the ad incentive program continued to be part of the CSC category management program, which was collecting data from over 3,500 stores. The CSC staff used this data to help retailers measure their performance against their competition and to make fact-based promotional recommendations and respond to current marketing, packaging, and pricing trends. To facilitate the CSC recommendations, a Pricing Study identified the opportunity to better understand optimum pricing leading to top retail performance in volume and total dollars. The analysis included both promoted and non-promoted pricing data for 1998, 1999, and 2000 from 5 major chains. The results were derived from peak season data, March 15th–May 31st, for each of those years and focused on performance of the one-pound container. Price range, promotion, discount, promotion type, event type, multiple items, and category performance identified optimal results. The key findings were:

1. Optimal one-pound non-promoted price to maximize strawberry volume and dollars ranges from $1.99–$2.59. Optimal non-promoted price average is $2.19.
2. Promotional discounts of 40% from no-promoted pricing deliver the best item, and category impact for both volume and dollars.
3. Multiple ad items and a rotation of ad items generate the best overall results.
4. Peak season promotional volume should range from 41%–80% of total volume sold, with an average of 63% sold on promotion.
5. Best results are achieved with frequent promotions at least every other week
6. "Buy One/Get One Free" (BOGO) promotions generate the highest increases in volume and dollars, more than a straight 50% discount.

These results provide the CSC category managers with a benchmark for assessing optimum pricing points for one-pound consumer pack strawberries and may vary by market and chain due to competitive factors and positioning. Gross margins are not incorporated in the study but it is suggested that retailers do the following:

1. Dismiss gross margin results if consumer price sensitivity is exceeded.
2. Recognize the importance of frequent strawberry ads to drive sales.
3. Use specific targets discounts from everyday price.
4. Use multiple ad items and vary the ad mix.
5. Appreciate the value of BOGO promotions.

One-pound consumer packs represented 77.6% of peak season sales. Lower labor costs because of easier handling, barcodes for more efficient pricing, and the demand for consumer ready packaging, has propelled this package, as well as half-trays at the store level. For the first time in 2000, the CSC, Kraft Cool Whipped Topping, and Sara Lee established a Cool Whip logo, which was featured on a one-pound consumer container while Kraft ran a national media campaign prominently featuring strawberries. Print advertising featured a strawberry *shortcut* recipe, and a national coupon offered free strawberries in major newspapers during May. Kraft also initiated a nationwide program with Pectin and strawberries for the May–July time period. The program focused on in-store merchandising of free strawberries with the purchase of Kraft brand pectin products. Freestanding displays included large photography of strawberry jam and recipe tear off pads. Category management also included retail-training programs on merchandising and handling techniques at store level. Collateral pieces also provide the trade with reference material and handling. Today, the CSC only provides POP material on request, while smaller regional chains, independents, major chains, and box stores rely on their own POP materials to enhance displays or none. CSC public relations activities were again important in circulating information on strawberries with newspapers, magazines, and electronic media. A satellite radio and TV tour featuring a registered dietician and *Prevention Magazine* editor Holly McCord touted the health benefits of eating strawberries. Due to late season volume resulting from expanded acreage of the Diamante variety, the CSC developed a late season radio campaign in four markets during a three week period to stimulate consumer demand and generate retail partnerships with advertising. The campaign ran as a test to evaluate retail performance and to determine the potential of a full-scale program for 2001. Reach and frequency were at a strong consumer level, and in addition to direct consumer advertising, the radio promotion provided the opportunity to partner with retailers in the specific markets. While the early results showed the radio advertising campaign to be a proven medium to drive short-term sales and generate consumer awareness, the cost to execute an effective program for 2001 was deemed prohibitive.

The public relation program consistently generated 500 million consumer impressions through print and electronic media. Food service was the recipient of much of the public relations effort. The research by Technomics International in the summer of 2000 showed high penetration of both fresh and frozen strawberries across all segments. Operators continued to use strawberries in all meal parts, with breakfast and lunch the most common for frozen. Frozen strawberry usage increased 7% in the lunch category since the 1996 study. National and large regional restaurant chains, along with distributors, were the primary target for promotional partnerships focusing on menu placement and increased strawberry volume. Constant communications with the USDA expanded frozen participation in the USDA school lunch program. Lower cost competition began to erode the export market in 2000. The goal continued to be maintaining market share through emphasis on quality and consistency, utilizing trade educational seminars, food service menu promotions, retail sampling programs, trade public relations, and trade visits. Trade newsletters and promotional materials were the key elements in the CSC's marketing program in Japan.

The 2,300 acre increase from 1999 caused a significant increase in volume and a much lower FOB price and reduced farm value, despite millions of dollars of CSC expenditures. Whether these numbers would have been lower without CSC programs is the question. We do know that the increase in the number of ads increased demand 2–3 times, but this was because of the grower/shipper, PSPROMO, cooperation with retailers.

In 2001, after 25 years, the CSC selected a new president, Roger Wasson, to replace Dave Riggs. This change clearly represented a change in direction, although it is unclear as to the specific reason and his tenure was brief. The format and content of the CSC Annual Report changed significantly from being a chronicle of past activities for the industry, trade, and media, to a report focused only on the strawberry industry. However, the CSC programs remained concentrated in the same areas and were directed toward collecting and disseminating category management data to retailers, foodservice operators, and industrial users. The CSC's ad incentive program remained the work horse of the marketing program. Table 8 indicates the annual increase in the frequency of strawberry retail advertising. In 1991 CSC tracked 440 retail chains, as mentioned earlier, with ads totaling 3,234 between March and September. In 2001, the number of retail chains decreased to 267 but the number of strawberry ads increased to 3,689. The number of ads increased per chain, as did the ads during the summer and fall months. In 2001 the acreage decline of 1,200 acres and the reduction in per acre yields from 27.6 to 25.2, resulted in a volume decrease and a substantial average price rise. The number of ads decreased slightly from 2000 to 2001, because of the volume declines. Category managers used all available information gathered over the years to make fact-based recommendations to retailers that responded to marketing and pricing trends. A major fact learned from 5 years of tracking was that consumers wanted specific packaging choices. The data indicated that strawberry sales could rise 17% when multiple packaging choices were offered. One-pound containers represented 81% of the annual category share. The half-trays and two-pound containers showed significant growth for summer sales and for larger families and special events. The 2000 Pricing Impact Study measures the impact of different pricing strategies at retail. The information was used during the peak 2001 season with retailer to help understand and develop pricing strategies to maximize strawberry performance. The study was built on previous category management research to identify retail price points that drive strawberry sales and profits. The incentive budget for incentive ad promotion was approximately $650,000 of a $1.5 million consumer/retail total. Although not an incentive ad program, the CSC began a new partnership with two highly recognized consumer brands to stimulate consumer demand through co-branded, direct consumer advertising. The strawberry industry responded with labeling strawberry one-pound containers in support of the program. While the program generated awareness for California strawberries through TV and print advertising, it was difficult to quantify actual increases in strawberry sales. Collateral, POP material, and training information continued to be important category management tools used to assist retailers with strawberry merchandising and handling techniques at store level. A new retail merchandising CD-ROM was developed to provide the trade with an electronic source in merchandising and handling information as the trade improved communication ability with individual stores. Retailers were given the opportunity to customize CSC marketing material for their use and the CD-ROM allowed the trade a cost-effective way to tap the CSC's resources.

Trade advertising and public relations have been the essential manner in which visibility of California strawberries has been maintained with consumers, retailers, food service operators and industrial users since the elimination of media in 1990. The CSC continued to run information-based ads during peak volume periods in such industry publications as *The Packer*, *Produce Merchandising*, and *Supermarket Business*. Since wide-reaching consumer advertising was not financially feasible, the CSC relied on public relations and in 2001 the CSC programs generated 500 million consumer impressions. Apparently, many expensive advertising campaigns do not match this number. Since the 1960s, the CSC has provided new recipes and nutritional information to newspaper and magazine writers and editors in order to keep strawberries prominently placed in publication throughout the US and Canada. Working with the American Dietetic Association and the Academy of Family Physicians, the CSC nutrition information was available to health conscious consumers. It also provided an easily accessible and user-friendly website and gave consumers 150 strawberry recipes, as well as information about the industry. In 2001, the CSC website received over 50,000 consumer visits.

The CSC goal for foodservice has been to expand fresh and frozen strawberry usage by securing menu inclusion. In order to accomplish this, new research has shown that chefs are constantly looking to other professional chefs and restaurant for new menu ideas. Therefore, in 2001 the CSC partnered with cooking professionals to create new and inventive ways to prepare California strawberries. An example is the CSC teaming with Disney World and hosted a culinary training session, which featured a presentation of ten new strawberry recipes personally created by celebrity chef Marc Valiani of Jianna's in San Francisco. The purpose was to demonstrate the endless usage opportunities, versatility and menuing potential for all types of Disney operation. CSC advertisements also highlighted ads showcasing recipes developed by celebrity chefs, appearing in publications targeting the casual chain and high volume independent restaurant. With accumulated information resulting from 1,000 leads and 1.1 million consumer impressions, direct mail featuring new recipes and usage ideas were sent to the top 400 casual chains and high volume independent restaurants. Frozen strawberries continued to be promoted in the schools, working with USDA/ Agricultural Marketing Service and USDA/Federal Marketing Service, to include strawberries in school lunch procurement. CSC export programs in Japan, Canada, Mexico and Hong Kong continued to receive funding from the USDA's Market Access Program (MAP), for use in trade communications, promotional materials, training and cooperative promotions to help increase fresh and frozen demand for strawberries. The CSC placed renewed emphasis on the quality assurance program and food safety response plans, including implementation of a food security plan.

During 2001, the CSC decided to assess various aspects of its operations as viewed by the CSC retailer, terminal market, and wholesaler accounts and conducted a retail and wholesale survey. The top ranked commodity group was the Washington Apple Commission, followed by the CSC. It moved up from number 5 in 1992 and number 3 in 1996, to number 2 in 2001. 55% of all respondents rated the CSC as extremely valuable or very valuable. The top 2 important services provided by the commodity boards were Advertising

Incentives and Food Safety Programs. Key retailers, key wholesalers, and smaller retailers and wholesalers rated advertising incentives as the most important service provided by commodity boards. Respondents were most satisfied with the CSC's POP materials and crises communications. The level of satisfaction with POP materials improved from number 11 in 1996 to number 1 in 2001, while the importance of POP materials continues to be ranked low. The reason appears to be the unwillingness of major chains to clutter their carefully designed stores with foreign material, which is labor intensive to set up and dispose. In 2001, crisis management dropped to the number 2 rank in services provided by the CSC, which may be due to the *cyclospora* outbreak that occurred slightly before the 1996 survey was conducted, while it was still number 1. In 1996 and 2001, promotional programs and materials offered knowledge of retail trends and remained the top two important aspects of personal contacts by the category managers. 61% of the respondents felt that more improvement was needed in the areas of retail promotion programs and 52% thought more personal contact was needed. A majority, or 52 %, of all respondents (key retailer and retailers), offered 3–4 different package types in their permanent assortment: 92% offered the 1-lb package, followed by 67% pints and 60% stems; 23% would add the 2-lb package to their permanent assortment; 39% of the respondents tried the 4 lb or half tray, in 2001; and 95% would add the 4 lb package to their permanent assortment. "Product shrink" was considered a major problem. Summer-time shrink: 61% of all respondents reported 0-10% and 47% reported 5-10%. Peak Shrink: 74% of all respondents reported 0-10% and 39% reported 0-5%. Arrival conditions accounted for 64% and temperature control management accounted for 55%, as the two key factors which contributed to shrink. The survey concluded that the majority of all respondents in all territories prefer commodity boards to communicate via e-mail. It is unclear from the survey whether the respondents no longer desire personal calls by category managers except perhaps for the incentive payments. The major barometers for program effectiveness continued to be the number of ads, specific promotions as mentioned above, but mostly proxies for sales such as consumer impressions and increased public relations on nutrition.

Demand Shift Variable: PSPROMO-CP, Private Sector Promotion and Contract Pricing

The incentive program portion of the CSCPROMO demand shift variable has been considered the work-horse of the marketing program, since wide-reaching consumer media advertising programs were not financially feasible. Additionally, the CSC has relied on public relations to raise the strawberry awareness of consumers, retailers, food service operators, and industrial users. We have previously traced CSC public relations through 40 years and the analysis of CSC incentive programs from their beginning in 1990. The incentive program seems to have been the *only direct measurable influence on demand* (See Table 8) with the exception of the media years when TV proved to be effective. Public relations remain difficult to measure, and only behavioral barometers measuring proxies for sales seem to be useable for evaluating board programs.

The very real problem with assuming that CSC incentive payments have impacted demand arises because of the very limited money contribution to retail advertisements. The promotions strategy used by the CSC (CSCPROMO) and shippers (PSPROMO) is different. Shippers must provide the supply of strawberries and the price to retailers for all transactions including an *ad special*. The timing and availability of strawberries for a retailer are both determined by the private sector and the role of the CSC is marginal at best. Their role has been to provide a minimal incentive payment of $1,000 to $1,500 for each ad over a certain base for their region, as well as nationally. The purpose of the incentive payment is to encourage additional retailer advertising by newspapers ads, publishing flyers, using in-store discounts, or "buy one get one free" promotions. Trade advertising, public relations, and personal contact by CSC category managers emphasized the supply conditions and availability of strawberries and contracted with retailers for additional ads over and above base levels. As an experienced shipper, it is my view that the information provided by category managers must come from shippers, who provide the timing, quantity and pricing of strawberries for all future ads. Additionally, each shipper serves as *category manager* for his own accounts and in many cases each retailer already has numerous shippers, or category managers, thus marginalizing this particular role of the CSC. As I noted earlier, retail ads cost $15,000–$20,000 per ad depending on the type and size of ad, and retail management's ad decisions are reflective of *all* category suppliers' input, including incentive payments. In this complicated retail decision-making process, CSC's incentive payment *may* tip the scale, but it appears rather doubtful. Many marketing boards are questioning whether incentive ads are mere additions to a chain's bottom line with little, if any, impact on ad decisions. It is interesting to note that the survey referred to above, mentioned advertising incentives as the most important service provided by commodity boards, and yet, as noted earlier, the majority of all respondents prefer that commodity boards communicate by e-mail, perhaps further minimizing their importance.

The private sector has the real burden of affecting demand. The concept of sales planning, or upward price management, is a strategy between individual strawberry shippers and retailers designed to reduce some of the risk in the volatile produce marketplace. In October, 1974 at the Produce Marketing Association annual meeting, I asked the Producers' Division members to *think retail*. I went on to explain that thinking retail requires advanced pricing, which can lead to retail advertising and promotion. "The sales planning concepts," I mentioned, "require each shipper, in conjunction with the CSC (or any trade association), to project industry volume, his own volume, and based on historical demand curves to project prices and volume for his retail customer promotion." This appeared to be the industry model until 1990 when CSC ad incentives began and category management was considered an integral part of the promotion process. In 2003, the ad incentive program was discontinued because the CSC apparently decided that the ad incentives were marginal in retail ad determination, but still important in providing competitive retail and industry information. In order to reduce the risk of the spot market, pre-committed pricing in advance of shipment and or contract pricing and all types of combinations were developed for the different market stages discussed earlier. The spot market has declined in importance as a result of major changes in the market, resulting from consolidation in the retail sector and the reduced number of suppliers. In 1999, the top 20 retailers accounted for a little over half of U.S. sales and sales of the top 4 chains rose from a 16% share of total grocery sales in 1992 to 29% in 1999. The top 8 retailers accounted for 25% of grocery sales in 1992 and 42% in 1999. The top 10 integrated wholesalers-retailers, including voluntary wholesalers like Supervalu, accounted for 50% of 1999 grocery sales.

In 1999, the total U.S. food system was $788.6 billion. Food retailing contributed $413.9 billion (excluding non-food, grocery store sales), which comprised 52% of the total, and foodservice contributed 374.4 billion, or 48% of the total. There are 30,600 supermarkets with 55.4% of retail food sales and 155,000 total stores selling food, including 83,000 convenience stores and around 1,000 supercenters that contribute 7.4% of retail food sales. The supercenter industry is growing rapidly with about $70 billion in 1999 and was expected to reach $180–$190 billion by 2003. Although the trend toward precommitted pricing and contracts began in the 1970s, the concentration and consolidation of retailers and the growth of supercenters and big-box warehouse stores have rapidly encouraged suppliers to enter into contracts, or precommitments, for quantities and prices including vendor managed automatic inventory replenishment. In 1999 Wal-Mart had over 721 supercenters, 1,081 discount stores and 463 Sam's Clubs, plus 1,008 international units of all types. Kmart had

102 Super Kmarts and 2,070 discount stores; Target had 16 Super Targets, 900 discount stores, and was adding 15 in 2000, with plans for an additional increase of 300 by 2010. Costco and Sam's Club probably handle at least $1 billion worth of produce from total sales of over $60 billion with huge growth predicted for this format. Sales planning/upside price management appeared to be the unorganized strawberry shipper strategy for dealing with some of the risk of volatile pricing; simultaneously "supply chain management" (Peters) was beginning to replace the traditional, fragmented daily sales orientation of the fresh produce business with partnerships focusing on year-end results. Traditional fragmentation on the supply side (growers and shippers) of the strawberry industry has yielded to significant consolidation; it became essential for competitive survival to diversify into every growing district in order to alleviate violent swings in regional weather and production declines. Failure of prices to increase relative to production cost increases has caused many small local growers to disband their farming operation, or become contract farmers, essentially working for the large, consolidated strawberry grower/shipper. The result has been 6 major grower/shippers with 70% of total volume beginning to cooperate with 6–10 major retailers and box-stores in some form of price and quantity system. *Could this be the end of the traditional, fragmented, daily sales orientation of the small, independent fresh produce business?* According to USDA research results, daily sales declined from 71% for a California commodity sample in 1994 to around 58% in 1999. A 2002 study by T.J. Richards and P.M. Patterson indicated that *contracts* in the produce industry have increased from 12% in 1969 to 36% in 2003. Their reason appears to be price risk reduction.

The industry survey that I conducted clearly indicated that vertically oriented grower/shippers have already begun moving toward year-round sourcing and category management approaches similar to the CSC category management model. My survey includes many variations of precommitments/contracts and differs from the macro data presented above. The nature of the strawberry industry and the 5 definite market time stages are the basis for differing precommitments and contracts, reflecting supply conditions, and the availability of fruit substitutes. It appears that contracts of a specific nature, with price and quantity established for specified time periods, are still rather limited, but gaining ground with Wal-Mart, other retailers, and in the foodservice industry. Examples of these contracts are:

1. Foodservice: This type contract would include the time period, 7 months beginning April 1 through October 31, estimated usage by month and peaking in June, and the contract price from April 1–14,

$10.75 FOB; April 15–May 12, $8.75; May 13–September 8, $6.75; Sept. 9–Oct. 31, $7.75. In the event of a severe shortage caused by an Act of God, such as an employee strike, natural disaster, or other incident that could not be foreseen, the agreement will be reevaluated. However, there is a cap on the increase in price, which may not exceed 150% of the contracted price. The price quoted are *lids*, the shipper will determine the price on any given day, agrees to supply the projected volume at the contract price, and the receiver agrees to use the amount projected.

2. This type contract involves a Mega Broker, who served as intermediary between the supercenter and the shipper: The contract begins on 3/13 and continues through 11/11. The shipper and Mega Broker are responsible for a designated group of supercenters, and to supply 100% of all supplies. Additionally, they will both offer *special buys* when the market allows. The price will be negotiated on a case by case basis, allowing the user to lower his price in the stores by cost averaging the contracted volume with the special buy purchase in order to move additional strawberries to the consumer. Regular season prices were established at $8.00 for 1 lb packages and $4.00 for half-trays. The quality standards specified no more than 15% total defects and only 8% of the total will be allowed for serious damage, of that only 3% will be allowed for decay. All products will be delivered with the PMA-UPC code numbers, packed daily, and loaded on standard size, reusable, CHEP pallets. It was also agreed that shipper and Mega broker would test the I.F.C.O. returnable plastic containers this season. An Act of God emergency clause was included and provided for the agreement to be reevaluated.

3. This contract is between the shipper and wholesale/retailer with the user's broker, negotiating with shipper for promotional volumes and pricing, in addition to the weekly contracted volume. The contract begins on April 1 and continues through Oct 12, for one truckload spread out over the week on various trucks, with product distribution to the destination user's branches to be determined by user's broker. The contract price will be $8.75 FOB, but any additional volume over the one load will be offered at the current week's market price. The Act of God clause in the contract provides for a cap on any price increase, not to exceed 150% of the contracted price.

4. This contract encompasses a 12-week period, April 1 through June 22, at which time local strawberries begin. The volume commitment is 2 truckload equivalents per week, spread over the week, to be distributed to the retailer's branches at the discretion of the retailer's broker. The contract price will be $7.90 FOB. Additional volume, over and above the 2 loads, will be offered at the current week's market price and in addition to the contracted volume; any promotional volumes and pricing shall be negotiable between the shipper and the broker for the retailer. The Act of God clause indicates that the price cannot exceed 150% of the contract price.

5. This contract begins on 4/2 through 10/20 and provides for 30–40 pallets per week at a price of $9.00 FOB for a 1 lb package ad pricing will be established once per month at $1.00 less than the above mentioned contract prices, with the promotion weeks to be agreed upon by user's broker. Promotions are to be confirmed 2 weeks prior to the pull date of each promotion. Special buys will be offered by shipper and buyer's broker when the market allows, and the price to be negotiated by shipper and broker on a case by case basis. An Act of God clause that places a cap on price increases not to exceed 150% of the contracted price.

6. A final foodservice contract begins in April and ends in Oct, with 2,000 to 4,000 trays per month, with usage determined by the buyer within the parameters. Any volume over and above the contract will be offered at the current week's market price. The contract is with the broker on behalf of the buyer and includes an Act of God clause similar to those above.

Both Patterson and Cook agree that about 40% of daily produce sales are on a contract or precommitted basis. Neither differentiates between the two, however my shipper survey indicates that only about 20% of daily volume sells on the spot market with the remaining 80% split between *contracts* at 15% to 25% and *precommitted prices* at 85% to 75%. The shipper-retailer contracts outlined above clearly prove their effect on price stability, but only partially explain the decrease in price variability indicated by Carter and Han (See Tables 9 through 15). The more important explanation is the domination of varying types of *shipper's verbal* precommitments in the produce market in general and the strawberry market in particular. Both these trends, contracts and precommitments, explain the relative ineffectiveness of the CSC and other marketing boards in directly influencing demand with minimum retail ad allowances and other category management techniques now utilized by grower/shippers.

As mentioned earlier, the 5 market stages have varying degrees of precommitments, while contracts, as evidenced above, cover the spring, summer, and fall markets. Although 6 major vertically integrated shippers control the strawberry market, there are numerous smaller shippers, lacking statewide production diversity, possessing only limited market power to compete with the major shippers in contracting or in making long term price and/or volume commitments. These shippers would be growing and shipping from one district in California with perhaps a small operation in other districts. An example would be a shipper who is only located in an early California shipping district and whose volume would be mostly in Stages I, January, February, and March, or until Easter. Farm value is at its highest, few substitutes are available before Easter, and demand elasticity has its lowest value in absolute terms. Demand will become more elastic later in the season. This category of shipper does not have contracts, although competitive pressures are requiring consideration. The spot market remains the system of daily sales, with only 25% of total sales at precommitted prices. Just prior to Easter, 50%–75% of total sales are precommitted, and after Easter in stage II, 75%–100%. It is important to note that these precommitments are not generally contracts with prices and quantities specified in writing, but merely verbal lids, subject to supply and demand changes. Committed prices will be forced lower if supplies are underestimated, but shippers are unable to raise prices regardless of supply shortage. This is the typical "one way street" market.

The large shippers and major chains that dominate the strawberry market represent the largest volume during Stages II and III, at least 50%–60% of annual total volume and farm value. These stages include the period between Easter and July 4, when demand is the most elastic and when precommitments are the most likely to affect the demand shift variable, PSPROMO. The shippers' survey indicates that major suppliers have already moved into year-round sourcing, streamlined distribution, and category management approaches. Their marketing programs are geared to reduce spot market reliance in order to reduce price risk. The increased cost of production, including escalating land rent, workers compensation, and other labor and production costs have left very little margin for error in pricing (See Table 16). Therefore all shippers are attempting to reduce price fluctuations with contracts and commitments, accepting fewer prices increases in order to prevent serious price reductions, which can or may accompany the spot market. Precommitments and cooperating with retailers and food service operators, in addition to contracts, are the methods of choice. The survey indicates that in Stage I, 25% of daily volume is precommitted, while 75% remains at the spot market in order to take advantage of price increases because of weather, and especially the possibility of bad weather in Florida, the chief competitive area in Stage I. During the peak market period, Stages II and III, from Easter until July 4, 65%–70% of total volume is sold at precommitted prices, with only 15%–20% on contract, many of which are in the food service sector. In Stages IV and V, during the summer and fall months with many competitive substitutes, 75%–85% of the volume is committed with price lids, but as in most price commitments, 70% of the prices are too high and are later adjusted lower at, or close to, the spot market price. As one shipper put it, "Upside price management reduces the magnitude of price fluctuations in the spot market, allowing for *fewer ups and downs* that are not as *deep or as long.*" The cause of this price adjustment process, even with commitments, is the *lack of perfect supply knowledge.* Supply predictions are about 80% accurate one week in advance, 3 weeks in advance the predictions are 50% correct, and the 4–5 week predictions are less than 50% accurate. Since the prediction error is generally an underestimation of volume, the result is a downward price adjustment from the committed price. The greater the magnitude of the underestimated projected volume, the larger the price adjustment and the longer time lag for market correction, with lower prices than justified by current supply and demand conditions. The main reason for the time lag in market adjustment is the time period between announcing the corrected FOB market and the ad/promotions price, and acceptance by the retailer. This time lag can be days or weeks, and the longer the lag, the lower the price.

Richards and Patterson, in their 2002 study on *Retail Contracting on Fresh Fruit,* failed to include the pervasive practice of precommitted pricing, which is the current marketing model, and also chose fruits for their study that have a storage capability and therefore more contracting than in the majority of the produce market which has unregulated daily supply and fluctuating prices. Additionally, they fail to include the large foodservice market, where contracts are more prevalent, more stable prices, and higher than the spot market. Their study also states that contracting prices are below the spot market, while my information is the exact opposite. Some shippers may contract a percentage of their volume at prices less than the cost of production, assuming that other contracts or precommitments will be an offset, or that the spot market will overcome income losses from the low contracts. Most shippers will try to maintain prices at cost or above, whether by contract (20%) or by precommitments (80%), hoping for fewer below cost sales and deep price declines. The study's "concern" that the "spot markets in produce have lost their role as the *locus* of price discovery," is illusory. The current practices of one large retail buyer, heavily involved in contracting as well as with special buys on the spot market, seem to have decreased the contractual percentage of the total usage at fixed prices and increased the spot market purchases and precommitments. One large shipper involved directly with the Wal-Mart designated warehouse program, which includes contract prices and volume for a specific warehouse and tied in with computer inventory control, has discontinued the program as have many strawberry shippers. The main reason for the shipper discontinuance is the inability to supply warehouse needs because of supply shortages and inability to cover the warehouse needs with proprietary volume. The mega broker has filled the gap between the shipper and this large buyer and taken the price and volume risk. This is a reversal of the contracting trend and a reliance on the spot market model with precommitments, guaranteed supplies, and price lids. The market power of this retailer and others, using the above model, assures supplies and suppliers at market prices or below. The shipper is assured of a market with fewer fluctuations and less prolonged market gluts. This model has been around since the 1970s, and the addition of retail contracting does not appear to have reduced the effectiveness of the market in facilitating the efficient allocation of industry resources, as the above study suggests. However, while chain retail buying power is increasing, partly expressed in the new demand for fees (slotting allowances), the huge physical volume they now procure also makes them even more dependent on shippers for stable, consistent, year round volumes. This may limit the exercise of market power in the fresh produce sector. As the food service sector grows and becomes even more powerful, they too are limited in their exercise of market power by their requirements for stable consistent volume, price and quality. The use of contracting has increased in the foodservice sector, but only seems to have replaced the lost retail contracting and left the total at 15%–20%. In this category as in retailing, the mega-broker representing a large group of independent operators is involved in the contracting of price and volume with shippers and processors.

The advertising history is the key barometer for measurement of the effectiveness of private sector ads and other independent use of category management tools to positively shift the demand curve. Another important private sector development is the use of multiple containers, which capture the multifaceted market of light and heavy users. The packaging revolution discussed earlier is another reason why retail advertising has expanded, increasing demand and decreasing price volatility. This packaging versatility has enabled the retailer to advertise more frequently with different containers. The CSC tracking data over the last 5 years indicated 1) consumers want specific packaging choices and 2) strawberry sales can rise 17% when multiple packaging choices are offered. The former retail practice of skipping weeks and substituting other commodities as an ad item has been altered by this multiple packaging, providing for an increase in the number and diversity of strawberry ads.

Trade ads, newspaper and magazine public relations information, and promotional material have always been a method for proprietary industry involvement. The expanded proprietary varietal development has provided another avenue for private category management, advertising, and promotion. One major shipper recently offered a 2005 VW Bug in a photo contest. The shipper offered to give the Bug to the family that looks the most photogenic while eating its branded strawberries. The contest was being promoted in the produce department of participating stores with displays of strawberries and point-of-sale material with the branded strawberries. Participants had to produce store receipts that showed the purchase of the branded strawberries. Their category managers deal with retailers and food service buyers directly and also company headquarters for special promotions, store openings, and other category information. Another major shipper's marketing manager conducted a National Strawberry Month and Mothers Day campaign in the shipper's major markets, providing TV and radio stations with branded press kits and other brand specific information. The sales manager and/or account managers were also the category managers responsible for data coordination and planning a strawberry program with price, volume, promotions, tie-ins, packaging, and product placement. The CSC category information is supplied by the Fresh Look Marketing Group and is the data base which includes weekly sales and monitors product movement through scanner cash register sales at 30,000 retail food stores. Feature ad data used by the CSC is provided from Promodata Leemis Services. All of this information is available to the private shipper's category managers, as well as retail category managers and top produce management. Therefore, this aspect of CSC category management appears to be less significant than other market information supplied to the strawberry shippers and growers by the CSC merchandisers. In the 2002 CSC Annual Report, which was previously sent to many audiences, including retailers, foodservice customers, and the media, the board chairman advised the reader that the report should only focus on one audience, our industry, and that the goal should be to provide growers with useful information about the focus and activities of the CSC. It was also decided that the merchandiser/category manager's role would be expanded to provide the same personal retail service to individual shippers. Thus the stage was set, for shippers to receive the same customer services provided to retailers. From the information provided by my survey, the category management data is interesting, dated, and not useful to either shipper or retailers, and has been available to any shipper/retailer for many years.

However, the role of the merchandisers remains important for the grower/shippers to continue providing weekly market information by major cities, on quality, pricing, merchandising and packaging trends, and the competition from other strawberry districts and commodities. Additionally, they are able to report on foodservice trends and promotions with national restaurant chains. In 2005, a new, *Strawberry Category Review* was developed on the CSC website as a tool to aid growers and shippers in their knowledge of the retail business. It is unclear what "new" means, since the reduction of merchandisers from 5 to 2 seems to indicate the relegation of the merchandiser/category manager to the other discarded programs, TV, radio, billboards, and retail incentive payments, rather than the development of a new program. Furthermore, fewer merchandisers limit the ability of the CSC to provide market information to shipper and grower CSC members.

Retailers in every market are aware of their competitors' advertising and pricing and therefore react according to their local, regional, or national, marketing strategy. It always appeared presumptuous for the CSC to assume that retailers needed information about how to manage their business. During the early stages of category management information, the data collected was conceivably more useful but, as all commodities became involved with category management, the phrase became a mere buzzword for CSC and other commodity board activities. This is perhaps one reason the majority of respondents to the CSC 2001 Retail and Wholesale Survey indicated that they preferred commodity boards communicate via e-mail, rather than by personal visit.

As the incentive payment program was discontinued and the CSC involvement in retail advertising ceased, the merchandiser's role in the public relations aspects of the new CSC "Red Edge" program, became dominant. The 2004 "Be well…Get The Red Edge" campaign was an integrated effort to gain consumer attention through magazines, newspapers, radio and TV releases, in-store promotions and food service programs. In addition to these activities, trade advertising and outreach at the American Dietetics Association (ADA), United/FMI and PMA conferences, and the Red Edge reached over the 325 million people in 2004. A few highlights included the cover of School Food Service and Nutrition featuring California strawberries and countless mentions in print and online media about strawberries being a *super food*, and healthy indulgence. The Red Edge nutritional message is part of national publicity about obesity and newly released dietary guidelines.

The preceding historical discussion of CSCPROMO and PSPROMO-CP from the beginning of CSC programs indicated the Board's annual evaluation of CSC effectiveness on the non-price demand-shift variables, where possible.

Unfortunately, the CSC's mostly discarded programs, such as TV, radio, billboards, tie-in promotions, display contests, infomercials, and food service restaurant promotions, have been somewhat measurable and evaluated by the Board on an annual basis. The demise of the CSC incentive program and other discarded programs, and the involuntary transfer of most category management functions to the private sector, have left the CSC with only programs involving generic trade, consumer, and foodservice promotion/advertising. The emphasis on health and nutrition is now the CSC major program highlighted by the slogan, "Get the Red Edge." This means that proxies for sales are the only measurable variables of CSC marketing programs that effect demand. The *Winter 2005 CSC Industry Newsletter* highlighted the new CSC mission, "Product research is the foundation of the 'Be Well, Get the Red Edge' nutrition program." The policies will expand nutrition research and knowledge about the health benefits of strawberries and manage the CSC industry-wide food safety program. The board also wants more production research; focus on grower issues such as soil disease controls and more knowledge about efficient production practices. Public relations are charged with managing the Red Edge message, understanding consumer values and concerns, developing media relations, and effectively communicating with CSC members. According to the newsletter, the CSC is now *knowledge-based* and *issues focused* to achieve what can not be done by an individual member or company. The CSC board must now decide on measurement barometers for assessing the effectiveness of their new direction.

The issue is more complex because the wearout factor (Kinnucan, Ward, Meyers, and Forker), discussed in the economic literature, becomes even more relevant because the health and nutrition issues have already been the subjects of CSC public relations for many years. There is evidence in the literature indicating that over time the effectiveness of a campaign may increase at first, but then decline due to the decease in advertising elasticity, which is attributed to "audience wearout." Additionally, almost every fresh fruit and vegetable, as well as packaged goods, are emphasizing health and nutrition and increasing the commodity *clutter.* Even if all of the barometers' measuring effectiveness of the Red Edge is positive, a correlation with demand will be difficult. The barometers of measuring effectiveness of proxies for sales will be those related to behavioral and attitudinal changes, such as changes in usage by heavy and light users, consumer and foodservice attitudes about strawberries, especially pertaining to health and nutrition, and other proxies mentioned earlier. Converting the change in attitudes into an increase in the non-price demand shift variable should be the CSC mission and a difficult one indeed, unless a *silver bullet* can be discovered by CSC nutrition research, which proves that strawberries are the solution to a major health problem. It is important to remember that the CSC objective, and also that of any commodity board, is to increase demand in order to increase grower profitability. The most likely answer to the industry's lack of profitability rests with the demand shift variables Price (P), Quality-taste (Q), private sector market and category management, whatever short run effect the CSC can provide with public relations, and the long-term effects of the *attempt* to change attitudes on health and nutrition. Finally, the next and concluding chapter will evaluate the CSC research/supply relationship and the effect on production, yields per acre, and grower profitability. The considerable reduction in the number of independent non-proprietary growers is indicative that grower profitability has decreased. Varietal improvement of taste and color and further increases in yields per acre seem to be the most likely methods of shifting the demand curve, decreasing costs per acre and increasing grower returns. The Red Edge mission statement, mentioned above, does not address the important variable, Quality-taste, or the important trend of pomology toward private varietal development by many major shippers, or the implication for the private nurseries, or the CSC-University of California relationship, as patent revenue declines.

Supply Shift Variable: APYCR, Acreage, Production, Yields per acre, Cost per acre, net Returns per acre

All of the CSC programs discussed thus far have pertained to the demand for strawberries as a function of many variables, including price and the non-price demand shift variables, which the CSC and the strawberry industry have attempted to influence. As we stated earlier the supply side of the model has been, and is, the main variable affecting price, volume, yields, production costs, grower's returns, and is the most affected by CSC expenditures. The cooperation between the CSC and the University of California has provided the model for agricultural research in every aspect of the production of strawberries, including the nurseries.

The uncontrollable seasonal characteristics of strawberries, including volume increases from January through March, peaking in April and May, causes prices to decline dramatically, and rise as volume declines. The volume available in the market explains a major portion of the variation in the FOB price, as well as the level. Therefore, CSC research in pomology and horticulture is vital to efficient production practices, which can increase yield per acre of quality strawberries at price levels related to total acreage and volume. A grower/investor in strawberry acreage must assume that the daily and annual projected FOB market price is determined by the historical demand curve and the projected total industry supply based on total acreage, varietal mix, and yields per acre. The data included in the 2004 report, *Sample Costs to Produce Strawberries* by the University of California Cooperative Extension,[1] provides the grower/investor with tables of historic prices, yields per acre, and grower net returns per acre, based on current total costs per acre in the 3 most important growing districts, Oxnard, Santa Maria, and Salinas-Watsonville and illustrates the small margins for these three areas (See Table 16, a summary of this study). The differing historical district prices are the result of specific supply and demand conditions in each of the 3 districts during each unique harvest period, depending on weather, early or late harvest, grower varietal mix, and the individual grower/district microclimate. An example would be the differences in price in Salinas and Watsonville, which are in close proximity.[2] The grower can only hope that the CSC can influence demand (shift the demand curve) through public relations and changing consumer attitudes and behavior toward strawberry usage, since other push and pull marketing tools have been practically discarded. The grower/shippers have the responsibility to influence demand with retail ads and the precommitments necessary to assure the level of pricing essential to market stability and prevention of disastrously low prices. However, per acre yields, costs, and quality are the variables under the grower's influence and control and determine net per acre returns. The data indicates the high financial, agronomic, and market risks that affect the profitability and economic viability of strawberry production.[3] The studies made every effort to model a production system based on typical real world practices and are the result of much consultation and advice from growers, who have verified its accuracy. The study reveals the extremely low margin of error for prices and/or yields lower than the averages assumed in the data. The following quotation referring to the year 2000 by a long-time strawberry grower, exemplifies the

---

[1] "Sample Costs to Produce Strawberries," by the University of California Cooperative Extension. (2004) South Coast Region (Oxnard): Yields and Returns, Table C, p.5; Net Returns Per Acre Above Total Cost, Table 4, p.17; South Coast Region (Santa Maria): Yields and Returns, Table C, p.6; Net Returns Per Acre Above Total Cost, Table 4, p.17Central Coast Region (Watsonville-Salinas): Yields and Returns, Table C, p.5; Net Returns Per Acre Above Total Cost, Table 4, p.15

[2] "Sample Costs to Produce Strawberries," by the University of California Cooperative Extension. (2004) Central Coast Section, Table C, p. 5.

[3] Ibid.

relationship of price and yields and the effect of drastic price or yield declines on the low profit margin:

"Obviously, the season was a disaster in spite of the highest yields in 6 years. Harvest costs remained constant over past years. Other costs increased. This is due primarily to the long crop cycle of the Diamante variety as compared to the Camarosa. The issue is whether or not the increased yields, about 15% to 20%, are enough to overcome the extra costs associated with the Diamante. Another concern with the Diamante is the issue of rot and albino. The berry was very weak on the Watsonville coast in 2000 and may have to be discontinued in that area. The situation with the berries is price driven at this point. It is not possible to correct growing costs and practices in such a way as to adjust for lower prices. The pricing level is so low there isn't enough room left to squeeze much out of the cost side. Furthermore, harvest labor costs have remained level for years and it is reasonable to expect the trend to continue. Due to the poor economics of strawberries, the acreage for 2001 was cut in half on conventional berries and organic was eliminated."

The quotation was written at the close of the 2000 season when the decrease in Diamante acreage began, especially in the Watsonville coastal climatic conditions. Production/acreage increased in the warmer, dryer Salinas area, although total acreage in the area declined for 2005, the only district to do so, except Orange County, where the decrease was due to land shortages.[4]

As indicated earlier, the CSC and most Commodity Boards have limited options available for influencing demand because of insufficient funds for extensive media campaigns, which in the past have been successful. The 6 largest grower/shippers must continue the marketing programs discussed earlier in order to better manage their varieties and quality, cooperate with retailers and foodservice operators for ads and promotions, and advance pricing based on supply projections, thus avoiding market gluts and low pricing. The CSC's future model should include public relations based on nutrition and category information for the consumer, retailers, and food service operators, and continuation of the proven successful research model, which concentrates on production, yields, costs, and quality/taste.

---

[4] Ibid.

Yields, quality, and costs are the variables under the grower's control; however the CSC has been and continues to be the research conduit for allocating funds to essential pomology and horticultural research projects. The CSC's research staff has the responsibility to act as the liaison between researchers and growers to oversee the entire research agenda. The CSC-university research model has existed since the early 1950s until the present time, and as Drs. Shaw and Larson together stated in the 2003–2004 CSC *Annual Production Research Report*, "The competitive position currently held by California strawberry growers can be traced to use of cultivars that have broad environmental adaptation, innovative production systems that maximize yields, fruit quality, and harvest efficiency, and use of pathogen-free plants, and soil environments." They recognize that, "California growers must overcome market and production cost barriers, while experiencing increased costs and decreasing fruit prices (in constant dollars), and the likelihood that a highly effective tool, methyl bromide, will be unavailable in the near future."

Illustrations on the following pages: Researchers, growers and others.

Picture 25. Top: Dr. Doug Shaw addressing growers at Watsonville test station
Center: TV interview of grower Kuni Shinta, Dr. Kirk D. Larson with food editors
Bottom: Larry Galper consults with, Dr. Doug Gubler, Dr. Greg Browne, Dr. S.M. Mircetich

Picture 26. Top: Daren Gee, John McPike, Dr. Doug Gubler, Dr. Greg Browne, Dr. S.M. Mircetich
Center: Dr. S.M. Mircetich, Dr. Greg Browne, Dr. Doug Gubler, Dr. Doug Shaw at test station for Watsonville Field Day
Bottom: Dr. Frank Westerlund with researchers, Ed Kurtz, Larry Galper, Cliff Golden

Picture 27. Top: Manabi Hirasaki, Peter Orr, Curt Gaines
Center: Dr. Kirk D. Larson, Experimental hoops at Irvine test station
Bottom: Raised bed fumigating in Santa Maria

Picture 28. Top: Victor Voth, left, and Herb Baum, right, with Roger Hamamura at his San Bernardino ranch, an example of Voth's preferred bed height.
Bottom: Roger's high gallonage sprayer, a Voth recommendation.

Picture 29. Top: Jack New
Center: Daren Gee, Tom Jones
Bottom: Ken Hasagawa, Mike Conroy
Packaging experiment: Bill Ito, 2nd from left, and Denny Donovan, far right, observe new crate.

Shaw's and Larson's research objectives are to breed improved short day and day neutral cultivars adapted to California climates and cultural environments as further detailed in the following excerpt from the study: "The criteria required for a successful California cultivar are constantly evolving. Currently, target traits for new cultivars in our program include: improved production attributes (marketable yield, production pattern, fruit size, ease of harvest, superior quality for fresh and processing markets, fruit appearance (color, flavor, firmness, shipping quality, shelf life), and resistance (or tolerance) to pests, pathogens, and the environmental factors or stresses. New cultivars must meet exacting minimum standards for all of the above traits, but the program currently emphasizes three target areas: 1) fruit quality, 2) harvest efficiency, 3) biotic (pests and diseases) and abiotic (weather and other environmental stresses) tolerance. Fruit quality has acquired an added premium in recent years due to increased production and market sophistication. Consequently, a major objective is the development of cultivars with the requisite quality attributes to enable expanded market share." To offset increased production costs, they have sought improvements in marketable yield and harvest efficiency (e.g. cull rate, consistency of fruit size and shape, plant size). Apparently, there is recognition that private breeding programs are growing and that proprietary varieties are increasing as a percentage of strawberry acreage (See Tables 2 through 5); however, they do not appear to appreciate the important negative taste/color aspect of the Ventana and Diamante varieties compared to the proprietary ones, since taste/color is not emphasized in their research report nor is the mention of brix sugar content measurement, a new spec by some retailers before purchase. For these reasons and more, the opportunity to compete with U.C. Davis for CSC pomology funding should be given to non-university pomology researchers.

Shaw and Larson continue to develop cultivars that are more tolerant of pests and disease, as well as weather and environmental stresses. For the past 9 years, research has helped develop strategies to *improve genetic resistance to foliar, fruit, and soil pathogens, as well as arthropod pests.* Their goal is to develop cultivars that are tolerant to important pathogens and that possess the requisite horticultural traits (productivity, production pattern, fruit quality, harvest ease, and cull rate).

Voth and Bringhurst developed the basic integrated cultural system, as outlined earlier, and the study recognizes that the basic principles on which these systems were founded are likely to exist in modified production systems of the future. However, Shaw and Larson state, "Many opportunities for improved cultural practices exist, although a continual concern remains that of soil fumigation. Research is continuing for the best substitute of methyl bromide, whether it be the use of bed or drip applied alternatives."

Drip applied material, Inline® (emulsified version of Telone C-35) and bed fumigants such as chloropicrin and Vapam® must be applied under polyethylene tarps. Adequate aeration of the root zone before planting is also essential for proper plant growth and vigor. "Weeds must be effectively managed to maintain high strawberry production, and strawberries are sensitive to weed competition, and weeds can harbor harmful diseases and insects," reports Dr. Steven A. Fennimore and Dr. Husein Ajwa, both of U.C. Davis, in the 2003–2004 CSC *Annual Production Research Report* quoted above. "Weed control in California has been accomplished through a combination of field selection, crop rotation, sanitation, hand weeding, mulching, preplant soil fumigation, and occasionally herbicides. Methyl bromide plus chloropicrin have long served as the foundation of strawberry pest control including weeds. According to Fennimore and Ajwa, "The phaseout of methyl bromide will leave growers dependent on fumigants such as chloropicrin, (Pic), 1, 3-D (Dichloropropene), plus Pic (Telone® C35, Inline®) and metam sodium." They further note, "Chloropicrin and 1, 3-D plus Pic are likely to be used extensively, but they are not always effective on weeds." This research was to evaluate weed control provided by the several alternatives indicating that Iodomethane plus Pic, 1, 3-D plus Pic, and Pic alone, were effective weed control treatments for strawberries, and the use of VIF tarp to replace standard tarp increased the efficiency of alternative fumigant applications. "Sequential applications of metam sodium or metam potassium in combination with the above, also improved weed control and lowered weeding times," states Fennimore and Ajwa. Dr. John M. Duniway, another U. C. Davis cooperator, researched the deployment of beneficial bacteria to improve root health, growth, and yield of strawberries and to improve the management of Verticillium wilt.

CSC recently invited qualified researchers to apply for pomology and regulatory/strategic issues which were two of their four priorities; the other two were entomology and plant pathology. The priorities for entomology were the control of two spotted mite, lygus bug, whitefly, thrips, and worms. Dr. Nick C. Toscano, Department of Entomology, U.C. Riverside, has found that broad-spectrum insecticides may control adult greenhouse whitefly, but only methonyl (Lannate®) was able to kill most adult whiteflies at relatively low rates. Dr. Frank Zalom, U.C. Davis, continued his diverse research into the control of most of the important insect and mite pests of strawberries, including spider mites, lygus bugs, greenhouse whiteflies, thrips, and Lepidoptera larvae. The plant pathology priorities were the control of foliar (Botrytis fruit rot, powdery mildew and anthracnose fruit/root rot) and root/soil, Phytophthora crown rot, and Verticillium crown rot. Dr. Greg Browne and Dr. Ravindra Bhat, U.C. Davis, focused on integrating the genetic, cultural, and control strategies for managing

Phytophthora crown rot (P cactorum) in strawberry. The results of their research confirmed that there are high levels of resistance to P cactorum in U.C. and private releases and found that the only effective control of crown rot was a plant dip treatment and that control was improved when spray or drip application of phosphonate were made in addition to the plant dip. Dr. Doug Gubler, U.C. Davis, evaluated the efficacy of various fungicide programs for control of powdery mildew, gray mold, and anthracnose fruit rot. He found that fungicides Quintec® and Pristine® provided the best control for powdery mildew, next was Switch®, with Rally® the least effective. There were no significant differences between treatments for Botrytis fruit rot, but all of the fungicide treatments had significantly less fruit rot than the untreated control.

Dr. Robert I. Krieger, Department of Entomology, U.C. Riverside, continued his work on measuring and mitigating pesticide exposure of strawberry workers. He has found that surface moisture can enhance the transfer of pesticide from plants to workers and therefore increase the chance for their absorption. Workers may be able to reduce their exposure to pesticide residues by wearing short sleeve shirts. This may prevent the absorption of pesticide residues that can be picked up by long sleeves from wet foliage. The effectiveness of a water rinse to remove trace surface residues was used to refute the claims for Fit ® (Proctor & Gamble's Fit Fruit & Vegetable Wash). Dr. Krieger claims this study resulted in the product's withdrawal from the market.

Dr. Frank N. Martin, USDA Agricultural Research Service (ARS), Salinas, CA. observed differences in plant growth and the timing of fruit production for chloropicrin followed by Vapam® or ® treatments. He also perfected techniques for nondestructive evaluation of plant growth by analysis of digital images and assessment of plant by measurement of reflectance of specific wavelengths of light from the plant canopy. These techniques will be used to evaluate the effect of pathogens, fungicide treatments and alternative soil fumigation strategies on plant health and productivity.

The CSC Research Committee identified six additional new research priorities for the 2006-2007 production season: plant nutrition and water use, cultural practices to improve the quality of nursery transplants, physiology of fruit development, cultural practices and technology to reduce labor costs, preserving the future availability of soil fumigants, and strawberry production without current fumigants. The CSC continued to express its basic, long held philosophy that *it will not fund research on topics or issues that individual farmers can do themselves* and that *proposed research should provide clear benefits to a majority of growers.* A complete evaluation of the 2003–2004 CSC-university research successes is found in the *CSC Annual Production Research Report.*

An interesting and potentially beneficial industry research project was the *protected culture trial* in Irvine, a 3-year study by Shaw and Larson, designed to discover if protected culture could benefit California growers by enhancing early season production with reductions in frost and rain damage. The tunnel system, almost 100% prevalent in the Mediterranean basin with a similar climate to California and wide use of U.C. cultivars, was utilized. *High tunnels,* utilizing steel posts and arcs covered with polyethylene film, were erected over the strawberry beds after planting and performance comparisons were made of early varieties with the use of different mulches, tunnel covers, and nitrogen fertility management programs to determine optimal cultural treatment for tunnel production systems in California. "Not surprisingly, the use of tunnels," state Shaw and Larson, "results in warmer soil and air temperatures, decreased PAR and UV light levels, and increased daytime and lower nighttime RH compared to outdoors." Early season tunnel yields have ranged from 114%, 156%, and 160% higher than outdoor plots with consistent fruit quality enhancement. "While the protected system may never be adopted by California strawberry growers, our project has examined this technology in considerable detail and we have developed a science-based technology 'package' that will enable growers to maximize use of this technology should they ever be interested," Shaw concluded. It is interesting to note that during the 3 year study, rainfall was below average, while 2005 was the second wettest in Los Angeles history.

The CSC-university research model was changed after the 1990 retirement of Voth-Bringhurst to include all of the scientists with relevant expertise in the 4 fields mentioned above, with separate budgets and specific research specifications. The CSC research committee adopts research priorities and the proposals are peer reviewed by qualified researchers as well as reviewed by the U C Strawberry Research Advisory Committee. Researchers must identify any additional funds that they anticipate receiving for their project and the CSC encourages proposals that involve cost sharing with other granting agencies or third parties.

An important element of the CSC-university research model is the propagation of *clean stock* (virus free), and the licensing of this disease-tested plant material to nurseries, in and out of the United States. The Foundation Plant Services (FPS), a self-supporting service department in the College of Agriculture and Environmental Sciences at U. C. Davis, performs the clean stock function. It produces, tests, maintains, and distributes premium, foundation-level virus and disease tested plant materials for use by nurseries licensed by the University of California Office of Technology Transfer (OTT). FPS strawberry custom services include:

1. Meristem tip culture, with heat treated strawberry cultivars
2. Maintenance of a strawberry selection at FPS as nuclear stock
3. Virus indexing of nuclear stock and of plants that are being introduced into FPS
4. The maintenance of the UC strawberry virus indicator plants
5. Virus indexing by use of graphing onto herbaceous indicators,
6. Virus Indexing By Elisa
7. The maintenance of all current UC commercially viable strawberry cultivars in a clean state and acting as plant repository
8. The development and use of genetic markers to identify UC strawberry cultivars throughout the world

The meristemming program began in the 1970s at U.C. Berkeley with Ruth Mullins directing the clean stock program. It was a major CSC research project, under the leadership of Bringhurst and Voth, and was transferred to Davis. FPS, in support of the U.C. Davis breeding program, maintains *a collection of current UC-patented cultivars*, while the OTT licenses these cultivars. The collection requires annual testing for virus diseases and genetic identity. Both plants and tissue culture meristem tip explants are distributed by FPS to UC licensees. In addition, FPS tests and maintains advanced breeders' selections produced by UC strawberry breeders, ensuring smooth release of disease tested propagating stock when new cultivars are introduced. Verification and maintenance of varietal purity and trueness to type is an important part of the strawberry program at FPS. In 2002, FPS was in the process of transitioning from *isoenzyme analysis technology* to *DNA fingerprinting* technology in order to monitor varietal trueness to type. Now in 2005, we see some of the DNA fingerprinting being completed a method that can correctly identify one strawberry cultivar from another strawberry cultivar with nearly absolute accuracy.

Many of the strawberry cultivars currently included in the FPS strawberry collection are U.C. patented cultivars and can be supplied only to those who are formally licensed by the OTT. When and if a prospective licensee (nursery) request cultivars, submits an *order* and is approved by the OTT, the FPS will then prepare the material for the licensee.

The OTT will permit the propagation of patented UC strawberry varieties in other countries if they provide and enforce Plant Breeders' Rights and other types of protection, such as honoring the protection of intellectual property rights. Two recognized international organizations are the International Union for the Protection of New Varieties of Plants (UPOV) and the European Union. UPOV is an intergovernmental organization, which cooperates in administrative matters with the World Intellectual Property Organization (WIPO), and has its headquarters in Geneva, Switzerland. The purpose of UPOV is to recognize and ensure an intellectual property right to the breeder of a new plant variety. The 40-member states of UPOV grant such a right to the breeder of a new plant variety. The European Union is the other international organization to provide and enforce Plant Breeders' Rights. Only Greece and Luxembourg of the European Union are not members of UPOV even China, with a history of complete disregard for intellectual property rights, is now a member of UPOV and has included strawberries as a protected plant patent in early 2005.

The OTT negotiates for Master Licensees in various parts of the world and with some nurseries separately where necessary. The Master Licensees are licensed by the U.C. Regents and granted exclusive sublicensing rights for specific cultivars in specific countries for propagation and transfer of plants for nursery or fruit production purposes. The Master Licensee has sublicenses in their assigned territory. While the propagation of patented strawberry cultivars is not allowed in unprotected countries by the Master Licensee's, the UC patented strawberry cultivars are sometimes allowed by the OTT to be used for fruit protection after ten years have elapsed from the time of patenting a strawberry cultivar in these unprotected countries. This allowance has been permitted by the OTT, due to pressure from California based strawberry nurseries, in order for the California based nurseries to compete with international strawberry nurseries that are illegally shipping UC patented strawberry plant material into these same unprotected countries. This practice will soon be eliminated due to an increase of needed police action by the OTT, and by police action by some of the assigned UC strawberry cultivar Master Licensee's that are doing business close to these countries or are responsible for the countries where the illegal plants are reported to have been generated.

The mission of the OTT is to ensure that the University of California's public service goal of providing its research for public use and benefit is implemented. This *technology transfer* is accomplished in many ways, for example, through educating students, publishing results of research and ensuring that inventions are developed into useful products in the commercial marketplace for public use. The university does not have the resources to develop products but, by pursuing patent protection for its technology, the university can offer a commercial company the exclusive rights to the technology and the incentive to invest in product development. To encourage and assist the university inventor in the use of the patent system in a manner equitable to all parties, a *patent policy* was adopted that: provides that the Regents obtain title to inventions or discoveries developed in the course of university employment, or with the use of university facilities funds under control of the university, and shares royalty income with the inventors. The purpose of the licensing system outlined above is to 1) provide a mechanism for transferring the results of university research to the public for public benefits, 2) meet obligations to research sponsors, 3) generate royalty income for the benefits of the university and the inventor, and 4) produce a profit for the university system and hence reduce the burden to the California taxpayer. It should be of interest to those that do pay California taxes, and to those that contribute to the CSC assessments, that the bulk of the revenue received presently by the UC concerning UC patented strawberry cultivars are from foreign sources and not from California strawberry growers.

The current strawberry patent program, as well as other UC patent programs, is in part the result of passage of the Bayh-Dole act of 1981, which provided a strong incentive for university-industry research collaboration. It was not until industry, academia and government recognized that their individual interests could be reconciled in the pursuit of commercialization, that the legislation was possible. Prior to 1981, the debate about patents by public universities was whether exclusive licenses would lead to monopolies and higher prices, whether taxpayers would get their fair share, whether foreign industry would benefit unduly, and whether ownership of inventions by a contractor is anti-competitive. Safeguards were included in the final bill and as the success of the Act became apparent; subsequent 1984 legislative initiatives and technical amendments broadened its reach farther. Colleges and universities immediately began to develop and strengthen their internal expertise to engage in the patenting and licensing of inventions, by developing teams with legal, business, and scientific background. During the 1970s and early 1980s, the strawberry patent program was very rudimentary and unprepared to enforce its patents. At one point during this period, I personally visited the grower in Florida who was rumored to have planted UC cultivars, and was asked why, at gunpoint, I was in his field. Later Bringhurst and Voth identified their variety in the grower's field and legal action was taken. Early patent enforcement was negligible, but as the patent program expanded so did enforcement and legal sophistication into the modern day OTT. However, there still appears to be a patent enforcement problem, since more acreage seems to be planted than can be accounted for. The current patent and licensing rules do not permit the use of plant parts, pollen or pistol, by any licensee or anyone else. Evidence of this new patent philosophy between public universities and commercial enterprises is reflected in the fact that the membership in the Association of University Technology Managers increased from 200 in 1990 to 2,179 in 1999.

Not only has the FPS and OTT been vital to the success and growth of the California Strawberry industry, but also the California Department of Food and Agriculture Strawberry Registration and Certification Program, Article 9 Regulations for California Certified Strawberry Plants, has been important in *Registration, certification, approval, and supervision of the California strawberry nursery industry*. Failure to comply with all state requirements shall be cause for refusal or cancellation of approval of plants as *foundation stock, registered stock*, or certification as *California certified* strawberry plants. The refusal to certify, or cancellation, will be determined when: the plant is off-type, or the plant, clone, or planting is virus infected, or the pest requirements have not been met. Reaction to indicator plants caused by unknown factors may also be cause to disqualify the specific foundation selection tested. The department issues official certification tags for foundation stock or registered stock, when the plants meet the technical requirements. My survey indicates that the certification process is antiquated and that the rules need to be much stricter for serious plant pathogens such as Colletotrichum acutatum, phytoplasmas, and others, which has not been diligently checked, perhaps because other countries are not as adamant about this problem.

The CSC-university research model, including the 4 major research areas outlined above, the FPS clean-stock system, the OTT efficient licensing and patent programs, and the CDFA Registration and Certification Program, has been and continues to be the foundation of increased strawberry yields, reduced or stable costs per acre, improved quality, and proper enforcement of patents. There is no margin for grower cultural, harvest, or varietal error, since reduced yields per acre, inferior quality (reduced demand), rising production costs, and competitive markets with declining strawberry prices in real terms, represent the *perfect storm* for lower net farm income. Perhaps, better OTT patent enforcing and more attention to proprietary competitive varieties would prevent a diminution in the long-time CSC-university partnership.

Supply Shift Demand Variable: FVIP, Freezer Volume, Inventory, Price

The processing strawberry sector of the California strawberry industry, PSAB referred to in Chapter 2, has been an integral part of the strawberry industry. Before the university varieties of 1945, the percentage of total annual volume was 70% freezer and the remaining 30% fresh was marketed only in California. By 1950, due to the new varieties and low freezer prices, the fresh market percentage had increased to 59%. After reaching a high of 63% in 1956, the freezer percentage steadily declined, averaging 35% to 40% until 1967, when the ratio became relatively stable at 70% fresh and 30% freezer. The 1962 introduction of the hybrid Tioga variety, with improved quality and yield plus the 10¢–15¢ per lb. freezer market, initiated the permanent reversal in the freezer-fresh ratio to a freezer-fresh market range of 25%–32% and 68%–75%, respectively. An average freezer price of 26¢–34¢ from 1990 to 2003 resulted in a steady 28%–32% freezer share of the strawberry market. The introduction of the hybrid varieties and the horticultural revolution caused the total acreage and production to increase tremendously and the freezer volume to expand from 140 million pounds in 1976 to 316 million pounds in 1990, 505 million pounds in 1999, 431 and 471 million in 2002 and 2003, at prices of 24¢, 30¢, 34¢, and 30¢, respectively. Meanwhile, the strawberry production from Oregon, Washington and Michigan continued to decline, unable to compete with California production, yields, prices, and the relatively arid climate and coastal weather influence where most California berries are grown. The following Tables 20 and 21 illustrate the decline in acreage, yields, and production in the above freezer producing states. Acreage in Michigan, Oregon, and Washington declined from 2,300, 7,800, and 2,400 in 1988 to 1,200, 2,600, and 1,700 in 2003, respectively.

Table 20
Strawberry Acreage and Production: 1988-1991

	Acreage				Yield Per Acre-Tons				Total Production-Pounds (000-Omitted)				Processing Tonnage (lbs) (000-Omitted)			
	1988	1989	1990	1991	1988	1989	1990	1991	1988	1989	1990	1991	1988	1989	1990	1991
United States																
Winter-Florida	5,000	5,300	5,400	5,500	12.8	13.0	10.8	12.0	125,000	137,800	116,600	132,000	-	-	-	-
Spring-California	17,650	19,900	20,000	20,500	24.4	21.6	24.5	27.0	860,109	857,928	982,897	1,105,141	203,732 (62.9%)	207,713 (75.3%)	316,385 (82.2%)	(6)331,924 (5)(85.2%)
Early Spring-Louisiana (4)	700	700	700	850	3.8	3.8	3.7	3.2	5,300	5,300	5,200	5,500				
Mid-Spring Group Total (1)(3)	6,000	5,900	5,800	5,700	1.5	1.5	1.5	1.5	18,000	17,500	17,000	17,000				
Late Spring-Michigan	2,300	2,400	2,400	2,100	2.6	2.4	3.0	3.1	12,700	11,700	14,300	13,000	4,000	4,000	4,000	3,600
Oregon	7,800	6,800	5,900	5,600	6.4	4.8	5.6	5.5	101,400	65,100	65,600	61,600	96,405	58,687	57,332	48,117
Washington	2,400	1,900	1,700	1,400	4.6	3.0	3.3	3.0	22,100	11,400	11,100	8,400	19,564	5,714	7,040	6,000
Other States (2)(3)	6,000	5,800	5,700	5,600	2.0	2.0	2.0	2.0	24,000	23,000	22,500	22,400				
Total – US	47,850	48,700	47,600	47,250	12.1	11.9	13.0	14.4	1,168,609	1,129,728	1,235,197	1,365,041	323,701	277,114	384,757	389,641
Less Processing Tonnage									(323,701)	(277,114)	(384,757)	(389,641)				
Balance for Fresh Shipping									844,908	852,614	850,440	975,400				
Add Fresh Imports: Mexico									35,116	30,671	27,723	(4)29,000				
Canada & others									(7) 4,282	5,422	4,393	(4)3,100				
Total Fresh Available									884,306	888,707	882,556	1,007,500				
Deduct Fresh Exports: Canada									(20,645)	(22,923)	(73,061)	(4)(74,000)				
Europe-Asia									(10,015)	(13,325)	(12,489)	(4)(12,000)				
Net Consumed in US-Fresh									853,646	852,459	797,006	921,500				

(1) Includes: Illinois, Missouri, Maryland, Virginia, North Carolina, Kentucky, Tennessee, Alabama, Arkansas, Oklahoma.
(2) Includes: Maine, Massachusetts, Connecticut, New York, New Jersey, Pennsylvania, Ohio, Indiana, Wisconsin.
(3) Estimated for certain states with small acreage. Crop Reporting Board disconnected service for these states.
(4) Estimated.
(5) Percent California to total.
(6) Includes Juice Berries
(7) Fresh imports in addition to Canada started 1988.

Source: Federal-State Market News
USDA Crop Reporting Board
Processing Strawberry Advisory Board

Table 21
Strawberry Acreage and Production: 2001-2004

	Acreage				Yield Per Acre-Tons				Total Production-Pounds (000-Omitted)				Processing Tonnage (lbs) (000-Omitted)			
	2001	2002	2003	2004	2001	2002	2003	2004	2001	2002	2003	2004	2001	2002	2003	2004
United States																
Winter-Florida	6,500	6,900	7,100	7,100	13.0	12.0	11.0	11.5	169,000	176,000	156,200	163,300	-	-	-	-
Spring-California	25,100	27,200	28,200	31,600	25.2	27.1	28.7	26.4	1,267,241	1,473,645	1,662,681	1,670,463	325,050 87.3%	402,849 89.8%	439,984 92.0%	(5)445,876 (4)92.4%
Early Spring-Louisiana																
Mid-Spring Group Total (1)	1,700	1,800	1,700	1,600	5.7	6.2	5.0	5.5	19,600	22,500	17,000	17,600				
Late Spring-Michigan	1,100	900	1,200	900	3.1	2.4	2.6	2.3	5,800	4,200	6,300	4,100	600	600	500	500
Oregon	3,100	3,000	2,600	2,400	5.5	5.8	5.8	6.7	48,200	33,800	29,500	32,400	35,718	31,534	27,530	26,292
Washington	1,600	1,800	1,700	1,900	5.0	4.5	4.5	4.0	16,000	16,200	15,300	15,200	10,883	13,791	10,290	10,492
Other States (2)	5,800	4,700	4,800	4,500	2.6	2.0	2.4	2.5	29,700	21,800	23,100	22,300	0	0	0	0
Total – US	44,900	46,300	47,300	50,000	17.2	18.9	21.4	19.3	1,547,541	1,748,145	1,910,081	1,925,363	372,251	448,774	478,304	482,660
Less Processing Tonnage									(372,251)	(448,774)	(478,304)	(482,660)				
Balance for Fresh Shipping–US									1,175,290	1,299,371	1,431,777	1,442,703				
Add Fresh Imports: Mexico,									68,829	87,410	88,560	(3)91,700				
Canada & Others									1,706	2,331	1,695	(3)1,504				
Total Fresh Available									1,245,825	1,389,112	1,522,032	(3)1,515,907				
Deduct Fresh Exports:																
Canada									98,404	121,983	146,644	(3)143,500				
Europe-Asia									29,444	34,584	47,753	(3)38,642				
Net Consumed in US-Fresh									1,117,977	1,232,545	1,327,635	1,333,765				

(1) Includes: North Carolina. Estimates discontinued for New Jersey in 2002
(2) Includes: New York, New Jersey, Pennsylvania, Ohio, Wisconsin. Virginia discontinued in 2001.
(3) Estimated for 2004
(4) Percent California to total.
(5) California only. Includes Juice Berries

Source: Federal-State Market News
USDA Crop Reporting Board
Processing Strawberry Advisory Board

During the same period, production declined from 12, 101, and 22 million pounds to 6, 30, and 15 million pounds. The cost of production is reflected in comparisons of the California yields and the other states. While California yields per acre exceeded 25 tons, Michigan, Oregon, and Washington yields per acre were only 2.6, 5.8, and 4.5 tons, mainly due to unfavorable weather conditions and varietal problems. Some Northwestern varieties command a premium price because of color and taste, but the combination of a price and low yields is generally insufficient for a profitable growing operation. The Northwest freezer production, which represented 40% of total US production in 1986, declined to 10% in 2003 because California's market share and total production had increased from 203 million pounds in 1988 to 440 million in 2003 thus making the Northwest a marginal factor in the processing strawberry sector.

In the earlier sections on CSCPROMO, the CSC marketing programs for foodservice and industrial users were outlined. Usage had increased, freezer prices fluctuated but remained relatively stable although declined in real terms, while per capita consumption of frozen strawberries had increased from 1.52 pounds in1975, 1.40 in 1985, 1.95 in 1995, and 2.03 pounds in 2003. The fact that per capita consumption had increased probably reflects greatly expanded production, but at what price and net return to the grower. The Achabal usage data indicates that only 5.3% of all households reported *usually* or *always* use frozen strawberries in 1992 and 1998. This maintains the decrease in frozen strawberry usage from 13% in 1982 to 7.6% in 1986. The shift towards more use of fresh versus frozen strawberries observed in 1992 was maintained in 1998.

In any given year, planting decisions and weather determine the quantities available in the fresh market. It is important to note that production in a given year is determined by the prior year plantings, since plantings are on an annual basis, perhaps 3–4 months before harvest in the following year. A late September or early October planting will begin producing in December of the same year, but more likely January in the following year, peaking in March and April, with a freezer harvest only after quality or market determined that the growers' switch from fresh market to the freezer. Shippers and growers cannot expand volume because of price, thus the quantity available in the fresh market is a predetermined variable, except for weather. Freezer prices *posted* early in the season for contract quantities can marginally influence the fresh supply, but it is only later in the season when the fresh market price has declined to comparable freezer levels, or because of quality, that fresh supply is reduced. However when contract quantity commitments are completed, freezer prices can be reduced or increased depending on the freezer market. In May of 2005, the fresh market had collapsed to freezer price levels, but because of rain in later growing districts, fresh prices advanced and growers in early districts shifted from freezer back to the fresh market.

Quality is an important determinant in this example, as is the contract commitment. There is a definite relationship between the fresh and freezer market, but the latter is a *salvage market* even when prices are equal. Growers will continue to ship to the fresh market, even those with freezer contracts, as long as fresh prices are above freezer equivalence, except when fresh quality has deteriorated. At this point, fresh and freezer prices are no longer relevant in determining which direction a grower will move. The fact is that per acre yields of fresh production is higher than freezer production and picking costs are lower. The reason for this is that fruit must be more mature for freezer harvest, requiring the fruit to remain on the vine 2–3 days longer than fruit for the fresh market. The plant fruit burden reduces continued fruiting and weakens future total production. Fresh harvesting, picking every 2–3 days, helps to retain a more productive plant for the entire production season. Additionally, higher yields per acre resulting from fresh, rather than freezer harvest, reduces harvest cost per lb or tray as well as fixed costs. Thus freezer prices must be substantially above fresh equivalence before a grower will switch from the fresh market to the freezer. In this sense, fresh market volume is a function of the freezer market under certain circumstances. The freezer market competes with the fresh market and the supply available for the fresh market will be reduced, if freezer prices are sufficiently high, thus influencing fresh market prices. Therefore, processor volume will have a positive relationship to the fresh FOB price. If freezer prices are near or below the cost of production, fresh volume could be higher and prices lower, especially if pre-season contract pricing is at this level. Processing prices also place a floor under the fresh market, but the floor can vary downward, changing the dynamics of the market place. For this reason, growers are reluctant to pick freezer in excess of their contracts or to switch too early in the season, due to the risk of lower prices and the inability to return to the fresh market because of quality problems, if prices are lower. New techniques for harvesting freezer berries have enabled an easier transition from fresh to freezer and vice-versa and relatively free of quality risk. This technique involves a simple tool that enables the picker to remove the calyx, which is mandatory for the processing plant (except when the calyx is acceptable for juice berries at much lower prices), even when the fruit is still in its fresh market stage. As a result, switching to freezer, when fresh market declines below the freezer equivalent, becomes instantaneous and market changes permit an immediate reversal to fresh market.

The freezer demand and prices are frequently a function of strawberry varieties that have superior quality, color, taste, and brix. These varieties *may* return a

4¢–6¢ premium over less acceptable varieties and affect the total varietal mix and fresh volume, up or down, in a given year. However, the 2005 intended higher freezer price for Camarosa over Ventana has not influenced the varietal mix because the increased early volume and total volume of the Ventana has caused a decrease in the total state acreage of Camarosa to 20% in 2005, from 40% in 2001 (See Tables 2 through 4). It is possible that the unpopularity of the Ventana in the freezer market, because of color and taste, could have an adverse affect on the fresh market with the decline of total freezer demand for strawberries However, quality and varieties, specifically the comparison between Diamante (the white fruit, as referred to by the industry) and Camarosa and Ventana (the red fruit), determine the freezer field prices. In 2005, the *red fruit* price was 28¢ per pound and *white fruit* was 26¢ per pound. Additionally, if the tool for cutting the calyx is used, an extra 2¢ is paid. Once again, the importance of quality is illustrated and the need for processor input in the pomology selection process is emphasized. As mentioned earlier, the Ventana is acceptable to the processor only if the Camarosa is unavailable.

The frozen export market has been and continues to be an important one, but there has been a significant reduction in the relative and total export volume. Only since 1999, export volume of frozen strawberries has drastically declined from almost 50 million pounds to 22.9 million pounds in 2003 and to 19.9 million pounds in 2004. This represents a drop from 12% of total frozen volume to 5.4%, and 5.1% in 1999, 2003, and 2004 respectively. The obvious reason for this decline is the increasing prominence of China as a producing and exporting country. From 1998 until 2004, Chinese freezer exports to the U.S rose from 1.2 million pounds to 5.1 in 2003, and 12.5 million pounds in 2004. The average total Chinese exports of frozen strawberries were 98 million pounds in 2001–2003, increasing to 147 million pounds in 2004. The Japanese market, which imported 63% of California frozen strawberries exports in 2003, decreased to a mere 25% in 2004. 30 million pounds were exported to Japan in 1999 and only 5 million in 2004. Canada became the largest user of California strawberries, importing 56% of California total exports, which was 10 million pounds in 2004, down from 17–18 million pounds on 1999.

A recent op-ed piece in the New York Times, quoted a headline in a Chinese newspaper, "China, Capital of the World." The rapid progress by the Chinese strawberry industry, with Chinese government subsidies and support, indicates the uncontrollable variables in the fresh and frozen markets. According to the CSC, "China is now the leading strawberry producer in the world with more than 1.8 billion metric tons produced in 2001–2003. More than 1 million individual growers produce strawberries on around 91,000 hectares (224,866 acres), of land." Much of the production is by very small growers with less than an acre, perhaps similar to the Louisiana berry deal up to 1960, with some growers planting and harvesting a few hundred plants. Most of the production is destined for the domestic market, where strawberries are sold in *wet markets* on street corners. The Chinese supply impact will be experienced in the world's processed market for the foreseeable future *only* until a sizable, sophisticated, domestic fresh market can be developed with all of the companion distribution and transportation systems. The standard of living also will need to be comparable to the US, an essential element for the consumption of a large volume from huge acreage, at prices sufficient to cover modern horticultural practices, including pesticides, fungicides, and other growing costs. Although the acreage is large, the potential for further Chinese inroads into the processed industry are significant, since they have numerous modern food processing plants including individual quick frozen (IQF), freeze dry, and cold storage facilities. The quality of Chinese products continues to improve and their pricing is very attractive because of low costs and their government's subsidies. China has extensive strawberry breeding and cultivation program and can produce many varieties, including Camarosa and the Oregon Totem variety. It is my understanding that China is already cultivating the Albion under a different name. The state-run universities are developing new varieties and are familiar with California horticulture. It is also important to note that China is still using methyl bromide and can continue to do so since they are not subject to the MeBr restrictions. Additionally, extensive greenhouse production exists in coastal areas utilizing modern horticultural systems and is able to supply quality berries for the fresh market on a year-around basis. An extensive system of super-highways has been developed and a state of the art world airport in Guangdong province, a major strawberry growing region near Hong Kong, features non-stop, wide body airfreight around the world. Other major Chinese cities, including Shanghai, Beijing, Dalian, and Shenyang have freight capabilities and are near major greenhouse production regions. The Chinese fresh market potential for exporting is developing, but success will depend on quality, cost of production and transportation and finally the price structure, which covers costs and/or the degree to which the government is willing to subsidize the strawberry industry in the short and long term.

In 2004, the CSC had a $762,000 budget for fresh and frozen export marketing. Continuation of the appropriate portion in the Japanese and Hong Kong markets would appear questionable since exports had declined: Fresh exports expanded from 1999 to 2003, and then fell 10% in 2004 from 16 million trays to 14 million trays, while total fresh production remained constant. This is a specific example of using the barometer (export volume and change) related to specific expenditures such as USDA's Market Access Program (MAP). MAP augments the CSC Export Marketing Program, targeting Canada, Japan, Mexico, and Hong Kong promotion activities and tries to position California as the most reliable source of high quality frozen strawberries produced under the most stringent sanitary conditions, compared to China. The focus continues on the foodservice sector, however according to industry interviews, the opportunities are quite limited and are based on the quality of California frozen product. While admitting the modernity of the Chinese processing plants, the CSC must emphasize the importance of multiple country suppliers, especially the US, which guarantee continuity of a quality product.

The impact of reduced exports on the demand shift variable, FVIP, total demand and price is yet to be determined. However, projected global freezer volume, current inventory size and composition (total availability) affect the freezer field price, which will finally be determined by the projected fresh market price and the break-even cost of production. Since acreage for a given year is a known variable, the freezer processors can assume the total volume and composition necessary to fill orders, the price level for early user supply contracts, and the supply-price risk involved in *playing* the spot freezer market when, as is inevitable, the fresh market and/or quality declines. Now that the processors are vertically integrated with growing, marketing, and processing, the operator can make these decisions based on internal and external economic considerations, knowing that his supply is determined internally without competitive interference or grower cooperation. Although it appears that there are many independent growers, 80%–90% of the total acreage is either vertically integrated or under contract with grower/managers. The economic analysis for determining fresh and freezer volume and price are the same, but the profit and loss for each activity is determined by the accounting procedures of each proprietor and the allocation of income and costs to growing, fresh marketing, and processing. There are approximately 15 processors of varying sizes with most of the volume originating in Santa Maria, Oxnard and Orange County with one processor at the California-Baja border to process strawberries from Baja California, Mexico. The non-integrated processors are mostly involved with growers on a contracted volume and price basis. The J.M. Smucker Co. is a grower/processor and also contracts for a specific volume and price with other growers, mostly for internal processing requirements. In 2005, Smucker closed their Salinas processing plant after 47 years. This was probably due to the ability to procure processed berries from China at much lower prices than domestic product.

For over 40 years, the CSC has encouraged the consumption of frozen strawberries by utilizing magazines and newspaper food sections, and public relations, as well as directly stimulating promotion by food service and industrial users. The introduction of yogurt and other end-use products has increased frozen usage, but per capita consumption has increased only 32% since 1965, while frozen production has risen from 70 million pounds to 476 million pounds in 2004. Since 1995, freezer production has risen 21% through 2004, while per capita consumption was lower. If per capita consumption and freezer field price were the barometers measuring the CSC program success, it would indicate that the program was a failure and should be discontinued. Perhaps one justification for continuance of these CSC programs is that most other commodity groups are doing the same thing, in the same way, and failure to continue *could* mean a loss of market. However, as we introduce behavioral barometers, the proxies for sales change between 1998 and 2003, according to a 2003 Consumer Survey, and indicate that attitudes have changed and there *may* have been a positive effect of public relations messages repeated to the foodservice and industrial users for many years. The results of the study are:

1. Consumers were more confident that California strawberries are safe to eat, increasing from 26% in 1998 to 37% in 2003.
2. Strawberries are, "A fruit I could eat every day," from 39%—1998 to 50%—2003
3. Strawberries are a nutritious snack for kids, 66%—1998 to 77%—2003.
4. Consumers understand that strawberries are a good source of fiber, 19%—1998 to 29%—2003.

It is unclear whether this consumer study applies to frozen berries, as well as fresh, and also how it compares to the study mentioned earlier by Achabal.

Commodity boards must decide their programs and budgets with the applicable barometers as measurements of success, scanty or nebulous as they may be. Proxies for sales may be the only barometer available to commodity boards.

Table 22
US Frozen Strawberry Pack, Imports, Consumption, Holdings, Average Field Price: 1950-1976

* California input does not include juice berries
** Grade No. 1, Cents per lb, FOB California Processing Plant

Source: USDA Crop Reporting Board, Federal-State Market News, Processing Strawberry Advisory Board

All figures in Million Pounds

Year	Calif*	Other US	Total US	Imports	Total Input	% Imports	Jan 1 Inventory	Total Available	Consump.	Dec 31 Inventory	% Total CA Crop	Average Field Pr**
1950	39.9	153.8	193.7	4.5	198.2	2.3%	47.5	245.7	148.0	97.7	41.3%	20.2¢
1951	40.2	117.7	157.9	6.3	164.2	3.8%	97.7	261.9	165.4	96.5	38.4%	20.0¢
1952	58.6	148.0	206.6	7.0	213.6	3.3%	96.6	310.2	204.3	105.9	40.7%	16.0¢
1953	93.7	133.9	227.6	8.1	235.7	3.4%	105.8	341.5	222.3	119.2	53.1%	16.5¢
1954	96.5	130.6	227.1	10.9	238.0	4.6%	119.2	357.2	247.4	109.8	55.4%	16.0¢
1955	119.5	156.7	276.2	12.0	288.2	4.2%	109.8	398.0	256.6	141.4	61.5%	17.0¢
1956	173.2	139.1	312.3	11.4	323.7	3.5%	141.4	465.1	269.3	195.8	63.8%	14.0¢
1957	118.3	142.6	260.9	13.8	274.7	5.0%	195.8	470.5	290.6	179.9	47.1%	10.4¢
1958	133.5	136.2	269.7	14.5	284.2	5.1%	179.9	464.4	296.4	167.7	52.8%	11.7¢
1959	82.9	165.3	248.2	14.4	262.6	5.5%	167.7	430.3	264.8	165.5	41.2%	14.0¢
1960	78.8	152.1	230.9	25.0	255.9	9.8%	165.5	421.4	260.3	161.1	43.1%	15.3¢
1961	80.1	141.6	221.7	29.8	251.5	11.9%	161.1	412.6	260.7	151.9	34.4%	11.0¢
1962	84.3	149.3	233.6	32.3	265.9	12.2%	151.9	417.8	258.8	159.0	34.8%	11.5¢
1963	94.2	138.9	233.1	34.5	267.6	12.9%	159.0	426.6	294.6	132.0	34.3%	11.5¢
1964	97.1	151.2	248.3	40.8	289.1	14.1%	132.0	421.1	263.8	157.3	38.2%	14.2¢
1965	75.9	112.2	188.1	53.9	242.0	22.3%	157.3	399.3	276.1	123.2	41.0%	16.9¢
1966	64.1	169.6	233.7	88.2	321.9	27.4%	123.2	445.1	290.4	154.7	33.5%	16.5¢
1967	67.9	142.1	210.0	77.9	287.9	27.1%	154.7	442.6	299.0	143.6	29.1%	14.7¢
1968	89.4	121.7	211.1	75.2	286.3	26.3%	143.6	429.9	283.0	146.9	25.9%	15.6¢
1969	75.4	104.8	180.2	92.3	272.5	33.9%	146.9	419.4	291.6	127.8	25.4%	16.3¢
1970	82.5	118.0	200.5	134.5	335.0	40.2%	127.8	462.8	296.7	166.1	25.5%	16.3¢
1971	76.4	120.9	197.3	84.6	281.9	30.0%	166.1	448.0	296.2	151.8	22.4%	14.0¢
1972	62.8	83.1	145.9	96.5	242.4	39.8%	151.8	394.2	289.8	104.4	20.6%	16.8¢
1973	95.7	74.0	169.7	133.9	303.6	44.1%	104.4	408.0	287.2	120.8	28.6%	20.2¢
1974	109.9	62.6	172.5	149.8	322.3	46.5%	120.8	443.1	294.9	148.2	27.5%	20.0¢
1975	118.8	70.2	189.0	109.6	298.6	36.7%	148.2	446.8	322.5	124.3	28.8%	19.2¢
1976	147.9	73.4	221.3	51.5	272.8	18.9%	124.3	397.1	293.2	103.9	33.2%	24.2¢

Table 23
US Frozen Strawberry Pack, Imports, Consumption, Holdings, Average Field Price: 1977–2004

Source: USDA Crop Reporting Board, Federal-State Market News, Processing Strawberry Advisory Board

All figures in Million Pounds

* California input does not include juice berries
** Grade No. 1, Cents per lb, FOB California Processing Plant

Year	Calif*	Other US	Total US	Imports	Total Input	% Imports	Jan 1 Inventory	Total Available	Consump.	Dec 31 Inventory	% Total CA Crop	Average Field Pr**
1977	180.9	50.4	231.3	92.3	323.6	28.5%	103.9	427.5	285.8	141.7	34.3%	22.0¢
1978	131.1	55.8	186.9	90.6	277.5	32.7%	141.7	419.2	310.5	108.7	25.9%	17.4¢
1979	148.8	61.6	210.4	128.7	339.1	38.0%	108.7	447.8	315.3	132.5	31.5%	27.7¢
1980	153.9	65.3	219.2	87.8	307.0	28.6%	132.5	439.5	287.6	151.9	31.3%	26.8¢
1981	136.3	65.7	202.0	62.0	264.0	23.5%	151.9	415.9	300.7	115.2	27.3%	28.8¢
1982	207.2	76.9	284.1	39.6	323.7	12.2%	115.1	438.8	300.1	138.7	36.2%	34.0¢
1983	185.5	101.8	287.3	49.9	337.2	14.8%	138.7	475.9	299.3	176.6	35.1%	34.1¢
1984	161.2	90.9	252.1	61.8	313.9	19.7%	176.6	490.5	324.5	166.0	22.4%	20.2¢
1985	181.1	73.7	254.8	52.7	307.5	17.1%	166.0	473.5	306.4	167.1	25.2%	20.4¢
1986	186.1	80.0	266.1	55.1	321.2	17.2%	167.1	488.3	341.7	146.6	26.9%	24.7¢
1987	245.8	127.2	373.0	96.6	469.6	20.6%	146.6	616.2	380.2	236.0	31.6%	28.7¢
1988	191.1	132.0	323.1	76.5	399.6	19.1%	236.0	635.6	400.4	235.2	23.8%	23.7¢
1989	191.7	74.9	266.6	52.9	319.5	16.6%	235.2	554.7	387.5	167.2	24.5%	24.2¢
1990	275.6	77.6	353.2	75.5	428.7	17.6%	167.2	595.9	397.6	198.3	32.2%	30.2¢
1991	317.5	69.0	386.5	77.4	463.9	15.7%	198.3	662.2	442.3	219.0	30.0%	27.0¢
1992	237.1	70.5	307.6	65.9	373.5	17.6%	219.0	593.4	419.6	173.8	26.0%	24.9¢
1993	339.6	73.9	413.5	66.9	480.4	13.9%	173.8	654.2	440.2	214.0	32.2%	27.8¢
1994	367.5	76.5	444.0	70.6	514.6	13.7%	214.0	728.6	483.9	244.7	32.4%	29.2¢
1995	341.9	65.1	407.0	92.1	499.1	13.5%	244.7	743.8	488.7	255.1	32.5%	26.5¢
1996	330.5	55.2	385.7	67.5	453.2	14.9%	255.1	708.3	496.3	212.0	28.1%	20.4¢
1997	337.7	59.4	397.1	71.1	468.2	15.2%	212.0	680.2	459.7	220.5	28.7%	28.2¢
1998	383.0	61.6	444.6	60.7	505.3	12.0%	220.5	725.8	524.4	201.4	34.0%	34.7¢
1999	423.3	51.7	475.0	100.0	575.0	17.4%	201.4	776.4	498.7	277.7	32.8%	34.0¢
2000	391.7	48.9	440.6	81.1	521.7	15.5%	277.7	799.4	488.9	310.5	28.0%	24.0¢
2001	334.4	51.9	386.3	74.6	460.9	16.2%	310.5	771.4	527.7	243.7	25.7%	30.6¢
2002	401.9	50.4	452.4	104.9	557.3	18.8%	243.7	801.0	534.6	266.4	27.6%	34.2¢
2003	387.0	42.1	429.1	105.2	534.3	19.7%	266.4	800.7	553.5	247.2	27.1%	30.2¢
2004	392.6	41.0	433.6	106.1	539.7	19.6%	247.2	786.9	482.3	304.7	26.7%	28.2¢

Table 24
US Per Capita Consumption of Fresh and Frozen Strawberries: 1962–2000

Year	Fresh	Frozen	Total Pounds
1962	1.60	1.25	2.85
1963	1.60	1.38	2.98
1964	1.60	1.31	2.91
1965	1.50	1.44	2.94
1966	1.40	1.49	2.89
1967	1.60	1.53	3.13
1968	1.80	1.43	3.23
1969	1.70	1.47	3.17
1970	1.80	1.48	3.28
1971	1.87	1.45	3.32
1972	1.68	1.41	3.09
1973	1.67	1.43	3.10
1974	1.80	1.44	3.24
1975	1.76	1.52	3.28
1976	1.75	1.42	3.17
1977	2.06	1.41	3.47
1978	2.17	1.49	3.66
1979	1.94	1.50	3.44
1980	1.96	1.37	3.33
1981	2.13	1.41	3.54
1982	2.27	1.46	3.73
1983	2.27	1.45	3.72
1984	2.92	1.45	4.37
1985	3.04	1.40	4.44
1986	2.96	1.58	4.54
1987	3.00	1.72	4.72
1988	3.46	1.74	5.20
1989	3.42	1.68	5.10
1990	3.12	1.84	4.96
1991	3.67	1.83	5.50
1992	3.64	1.72	5.36
1993	3.60	1.77	5.37
1994	3.98	1.63	5.61
1995	3.84	1.95	5.79
1996	4.13	1.92	6.05
1997	3.96	1.73	5.69
1998	3.83	2.14	5.97
1999	4.35	1.98	6.33
2000	4.49	1.77	6.26
2001	3.92	1.79	5.71
2002	4.27	1.94	6.21

Summary and Conclusions

Stephen Wilhelm and James E. Sagen, authors of the previous book, *History of the Strawberry*, provided considerable information about strawberries in general and the California industry as an appendix. Unfortunately, they excluded the *revolutionary* contribution of Dr. Royce Bringhurst and Victor Voth, before the publication of the book in 1972 and obviously afterwards. Also neglected was the discussion of the competitive advantage derived from the Thomas-Goldsmith rumored transfer of plant materials from the pomology department at University of California, Davis, to Strawberry Associates/Driscoll Inc., a proprietary breeder, grower, and shipper. The everbearer advantage that resulted in the Driscoll industry domination may have been the effect of university cultivars, selected during the pomology turbulence at the university. Ostensibly, an inventory of university cultivars was never taken and the detailed cultivars transfer of plants by Thomas-Goldsmith is unknown by university officials to this day. It would appear that the Driscoll competitive advantage gained from the transfer of knowledge from Thomas and Goldsmith, former University of California employees, was in all probability the cause of their competitive advantage and continued industry dominance. Such preeminence does not currently exist in any other area of California agriculture, because of the likelihood of non-competitive agreements for University of California employees in sensitive, creative areas involving patents.

From the above university confusion, Victor Voth, and shortly thereafter Dr Royce Bringhurst, transformed a troubled and disrupted industry into world dominance. Almost all rules of horticulture were excluded and new ones substituted. Voth first developed sprinkler irrigation to replace the long-standing furrow irrigation system for salinity control and continued with all of the other horticultural revolutions mentioned earlier. Growers, who so closely watched and learned from their mentor, implemented the horticultural breakthroughs in their fields before Voth could even publish the research results. Other developments such as summer planting, mulching, annual planting, winter planting, nursery elevations and plant chilling requirements, were vigorously implemented by a dedicated Victor Voth. During this period, other university scientists were working on Verticillium wilt, which caused the 1950's disastrous industry decline. Wilhelm's development of methods for soil fumigation, using chloropicrin and methyl bromide and then applying them under plastic tarpaulin to hold materials in the soil longer, was the necessary system for Voth's horticultural successes. Many other scientists from Berkeley, Davis, and Riverside worked closely with Voth and Dr. Bringhurst. Entomologists and plant pathologists were also researching problems created by mites, nematodes, and aphids, while other diseases than Verticillium wilt were addressed. Meanwhile the partnership of Bringhurst and Voth was developing the successful breeding program that changed the strawberry landscape for many years to come. Voth's ability to communicate with the growers to gain their trust, cooperation, and friendship, distinguishes him from the others in a sea of strawberry scientists and qualifies him as the Strawberry Man from La Mancha. The CSC-university research model in the final chapter offers a blueprint for commodity board research programs.

The research model and the resulting programs are the responsibility of the CSC, financed and directed by growers, shippers and processors under the California Department of Agriculture. Additionally, the CSC has the responsibility for efficient development and management of a national and international advertising and promotion program, which will enhance the competitiveness of the California strawberry industry within the national and international marketplace. I have chronicled the CSC advertising/promotion programs during its history, as well as all CSC research activities, with the objective of providing the framework for analyzing and evaluating these activities in a descriptive model, in order to justify the mandatory check off or assessment.

In addition to updating the *History of the Strawberry*, the purpose of this book is to analyze the relationship between the demand for fresh and frozen strawberries and the variables affecting that demand. The question is whether CSC research and marketing programs have caused demand to increase and the answer requires the development of a descriptive model constructed to provide information necessary for understanding the strawberry market and to describe the basic features and internal functioning of the industry. The model described in this book is used to explain the relationships between the variables and the CSC interaction with grower/shippers, their separate and joint affect on these variables and ultimately demand, price, and net farm income. I am suggesting that the quantity demanded, Qt, depends upon the corresponding Price of Strawberries, PS; Quality, (Q), CSC marketing activities, CSCPROMO, Private marketing programs, PRPROMO-CP, Freezer Volume, Pricing and Inventory, FVPI, the Price of Substitute Fruit, PSF, or:

$$Qt = f (PS, Q, CSPROMO, PRPROMO, CP, FVPI, PSF)$$

This model is an analytical procedure for evaluating alternative industry plans for any marketing board whose goal is to improve production and marketing. Most econometric simulations have been difficult for the marketing boards to comprehend (Pfouts). Also, the literature involves the econometric analysis of the effects of generic advertising (not promotion) on sales and does not include market analysis suggested by most economists and the writer (Johnson). The reason for including market analysis is the need for some tangible asset other than sales and advertising expenditures, the traditional indicator of advertising effort, for measuring the effects of advertising (Blaylock). The market information and analysis derived from the barometers have enabled me to evaluate the variables affected by the CSC and the private sector thus providing an understandable tangible asset.

The market volume explains a major portion of the variation in FOB prices for fresh and processed berries (Chalfant and Carter). FOB prices tend to follow an annual pattern, falling when supply increases. Seasonal variations are evident as volume increases in January through May, when prices decline dramatically and rise in a similarly drastic manner late in the season as production phases out. The uncontrollable, strong seasonal characteristics of supply cause most marketing problems. The simple correlation for the two series is -53 indicating that, even in weekly data, the volume available in the market explains a major portion of the variation in the FOB price (Han). All marketing boards should understand this relationship before embarking on programs to influence demand.

Once it is understood that the volume explains a major portion of the variation in price, and that demand is mostly a function of price, then other demand shift variables can be evaluated. Positive measurable performance outcomes were reflected in demand increases from high levels of TV, and relatively insignificant for low weight TV, radio, and billboard ads. Until 1990, when incentive chain advertising payments began, public relations campaigns involving health claims, news releases, recipes, etc. and the media efforts mentioned above, excluding network TV, seemed to influence the consumer from 1982–1998, as indicated by the proxies for sales barometers of the Achabal studies on pages 65–68. However real prices declined, unable to keep up with inflation, while strawberry crop value rose because of productivity gains (See following Tables 27 & 28). Grower prices for fresh strawberries increased an average of 4% per year since 1970 (Tables 25

&26), while processed prices rose 2% per year (Table 30). Real growers' fresh prices (inflation adjusted) declined 36% from 1970 to 1993, and processing prices slid by 57% (Tables 27 & 28). Prices would have been 87¢ per pound (actual 24¢ to 60¢) and processing prices 47¢ (actual 16.3¢ to 27.8¢), if prices had kept up with inflation. The trend of inflation adjusted retail prices from 1980 to 1993 was nearly flat, increasing only 4% in the 14 years, although fresh retail prices rose about 5% annually, or 75% during the period (Table 29). It is also important to note that the increased cost of packaging, transportation, and marketing rose faster than grower's prices. It would appear that the CSC marketing programs except for heavy weighted TV, have not resulted in positive measurable performance outcomes for growers (real prices and farm income), that justify mandatory assessments in the future. In order to further defend CSC marketing programs, it would have to be argued that proxies for sales, represented by attitude and behavioral changes, caused prices and farm income to be higher than without the programs that required the mandatory assessment. It would be a stretch to conclude this.

Table 25
Grower Prices for Strawberries, Cents Per Pound: 1970–1993

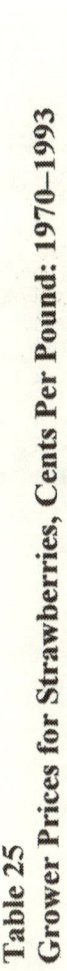

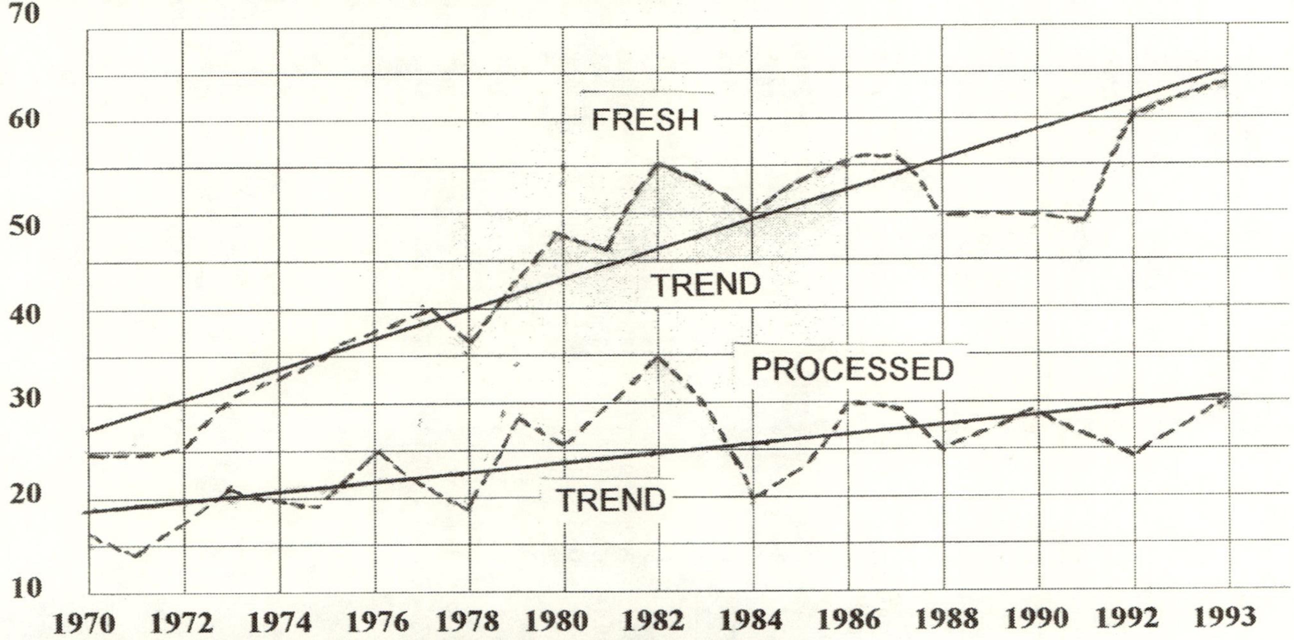

Source: USDA Economic Research Service, Diane Bertelson

Table 26
US Strawberry Production, Utilization, Prices and Values: 1970-1993

Year	Utilized Production			Grower Prices			Value of Production		
	Fresh	Freezer	Total	Fresh	Freezer	Total	Fresh	Freezer	Total
1970	316.4	179.6	496.0	24.8	15.6	21.5	78.5	28.0	106.6
1971	340.4	180.3	520.7	25.4	14.5	22.5	86.5	26.1	117.2
1972	321.1	139.1	460.2	27.1	16.8	24.0	87.0	23.4	110.4
1973	316.4	163.2	479.6	31.0	21.0	27.6	98.1	34.3	132.4
1974	370.6	168.2	538.8	32.5	20.7	28.8	120.4	34.8	155.2
1975	377.4	173.2	550.6	35.5	19.9	30.6	134.0	34.5	168.5
1976	369.5	211.2	580.7	37.7	24.5	32.9	139.3	51.7	191.1
1977	429.8	232.1	661.9	39.1	22.4	33.2	168.1	52.0	219.8
1978	477.9	181.3	659.2	36.7	18.8	31.7	175.4	34.1	209.0
1979	436.0	202.3	638.3	43.4	28.5	38.7	189.2	57.7	247.0
1980	482.1	219.6	701.7	47.9	26.3	41.2	230.9	57.8	289.1
1981	537.5	204.1	741.6	47.1	28.3	42.0	253.2	57.8	311.5
1982	589.6	293.4	883.0	55.2	33.8	48.1	325.5	99.2	424.7
1983	585.4	308.1	893.5	53.0	31.5	45.6	310.3	97.1	407.4
1984	748.2	242.7	990.9	49.0	19.3	41.7	366.6	46.8	413.2
1985	754.1	264.7	1,018.8	52.6	20.4	44.3	396.7	54.0	451.3
1986	734.8	284.5	1,019.3	57.6	28.4	49.4	423.2	80.8	503.5
1987	780.4	336.9	1117.3	58.5	28.5	49.4	456.5	96.0	551.9
1988	855.5	323.6	1179.1	54.1	25.2	46.2	462.8	81.5	544.7
1989	861.6	280.4	1142.0	53.9	26.1	47.1	464.4	73.2	537.9
1990	864.2	390.1	1254.3	55.3	28.7	47.1	477.9	112.0	590.8
1991	971.5	397.4	1368.9	54.1	27.4	46.3	525.6	108.9	633.8
1992	980.3	335.1	1315.4	61.5	24.7	52.1	602.9	82.8	685.3
1993	987.6	436.2	1423.8	63.1	28.4	52.5	623.2	123.0	747.5

Table 27
Grower Prices for Fresh Market Strawberries, Constant Dollar, Cents per Pound: 1970–1993

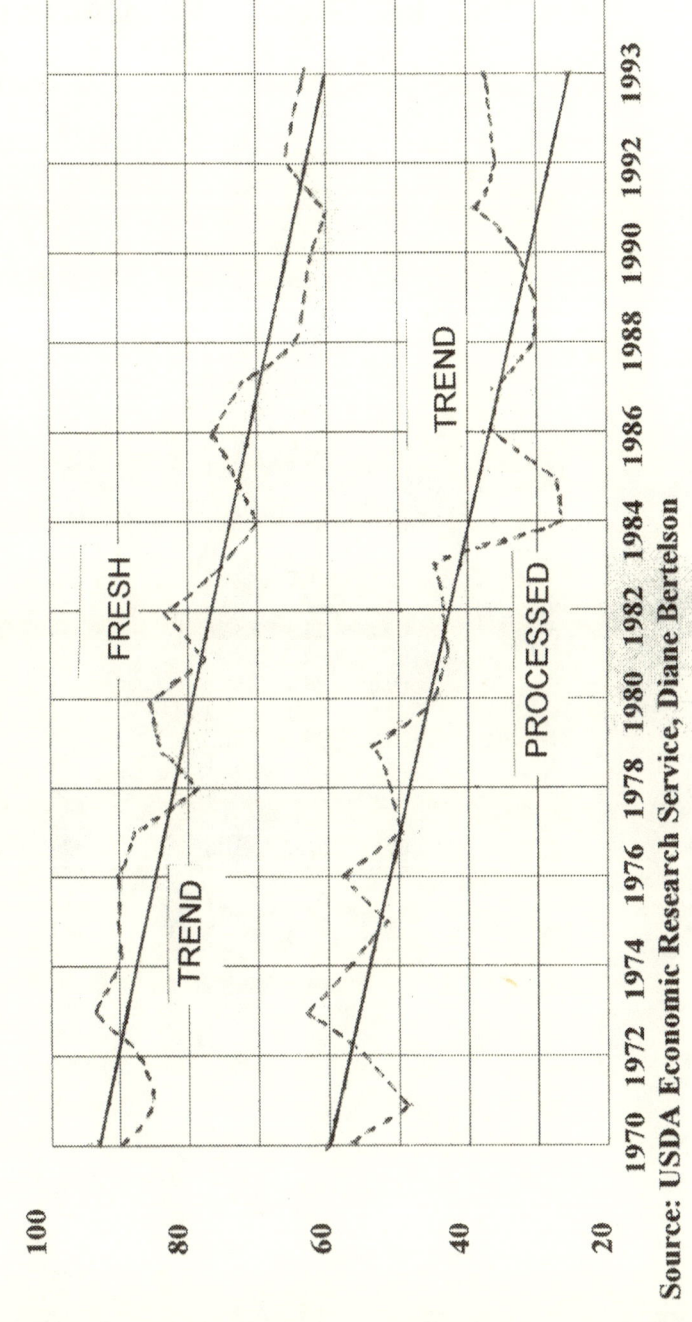

Source: USDA Economic Research Service, Diane Bertelson

Table 28
Real US Strawberry Prices and Values: 1970–1993

Year	GDP Implicit Price Index	Real Grower Prices Fresh Freezer Total			Real Value of Prod. Fresh Freezer Total		
	1993=100	Cents per Pound			Million Dollars		
1970	28	87.5	55.1	75.9	277	99	376
1971	30	85.1	48.6	75.3	290	88	392
1972	31	86.3	53.5	76.4	277	74	352
1973	33	93.0	63.0	82.8	294	103	397
1974	36	89.7	57.1	79.5	332	96	428
1975	40	89.4	50.1	77.1	337	87	424
1976	42	89.3	58.0	77.9	330	123	453
1977	45	86.7	49.6	73.6	372	115	487
1978	49	75.4	38.6	65.1	360	70	429
1979	53	82.0	53.8	73.1	357	109	467
1980	58	82.8	45.4	71.2	399	100	500
1981	64	74.0	44.4	66.0	398	91	489
1982	68	81.6	50.0	71.1	481	147	628
1983	70	75.3	44.8	64.8	441	138	579
1984	73	66.7	26.3	56.8	499	64	563
1985	76	69.0	26.8	58.1	521	71	592
1986	78	73.6	36.3	63.2	541	103	644
1987	81	72.5	35.3	61.2	566	119	684
1988	84	64.5	30.1	55.1	552	97	650
1989	88	61.6	29.8	53.8	530	84	614
1990	91	60.5	31.4	51.6	523	123	647
1991	95	56.9	28.8	48.7	553	115	667
1992	97	63.1	25.3	53.4	618	85	703
1993	100	63.1	28.4	52.5	623	124	747

Table 29
Retail Prices for Fresh Market Strawberries,
Constant and Current Dollar Per Pound: 1980–1993

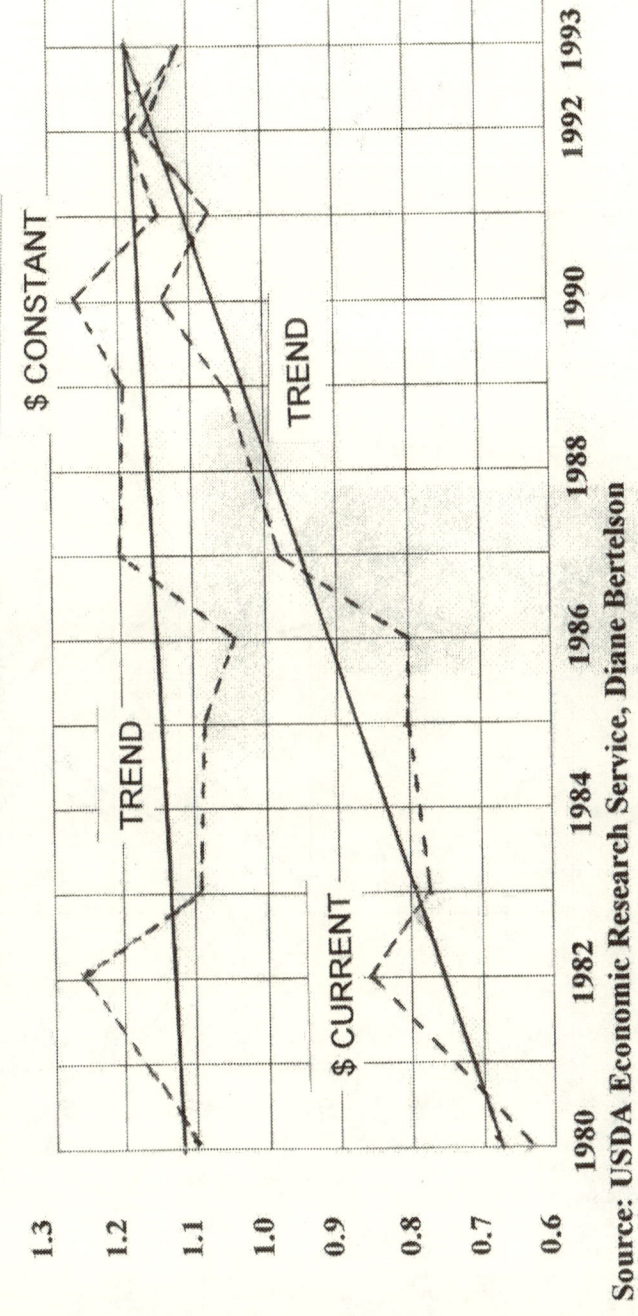

Source: USDA Economic Research Service, Diane Bertelson

Table 30
US Processed Strawberry Prices, Cents Per Pound: 1980–1993

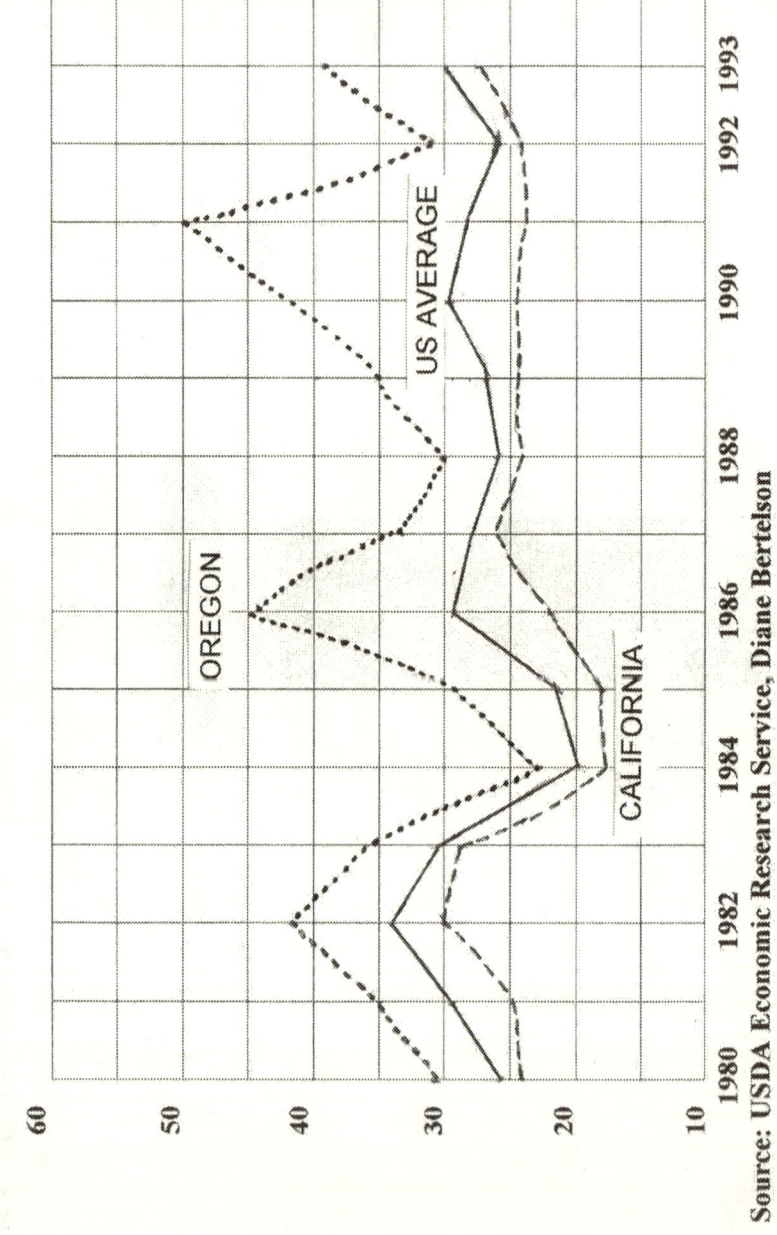

Source: USDA Economic Research Service, Diane Bertelson

Quality plays a vital role in demand and price determination and is the most important demand-shift variable. The CSC-university model, which utilizes many university researchers for horticulture and UC Davis for pomology, has been and continues to be the successful vehicle for improving quality and increasing yields. Quality, taste and. color have been the most important elements in consumer decision making when choosing alternative berry varieties, which frequently is reflected in proprietary brand preference. The price differential from 1960 to 1983 was $2–$3, because of the quality differential on arrival at the final shipping destination. Before the Selva variety in 1983, the non-proprietary varieties could not be shipped from June through November, leaving the processor as the only important user of these varieties during the summer and fall. The Selva provided non-proprietary growers the opportunity to compete during the summer and fall months, but at substantial price discounts. In 2000, the Diamante provided industry competition with proprietary varieties, but the color and taste were still inferior and disliked by processors. However, the firmness and tremendous yield per acre have overcome some of the competitive taste disadvantage, and the long period of trade dissatisfaction with non-proprietary varieties. Price differentials can still be significant, especially later in the season. USDA market data indicates a $1–$3 price difference, depending on appearance, size, and condition. In 2005, processors are paying 2¢–4¢ less per pound for the Diamante variety than any other variety because of color, which may be related to sugar content. Currently, a global user of California juice quality berries pays 8¢ per lb. for brix of 6.0–6.9, 10¢ on brix of 7.0–7.9, and an additional premium of 1¢ for brix exceeding 8.0. The new university variety, Albion, may lessen the competitive gap because of appearance and taste. Quality is an important and measurable demand shift variable that is greatly affected by CSC-university policy and is directly related to the fresh and freezer price.

There is no doubt that a CSC mandatory assessment rate is justified to provide the income necessary to fund research in pomology and horticulture. The resultant increased production, yield per acre (Tables 6 & 7), and improved quality (demand) have more than offset the drop in prices (Tables 27 & 28), causing an increase in the value of farm production (Bertelson). It is unlikely that the current 4.5¢ rate would be necessary, if the CSC marketing functions were reduced or eliminated altogether, as suggested earlier. In order to assess pomology, horticulture, and marketing functions based on cost and effectiveness, the separation of functions would be required, including a separate mandatory assessment rate established for pomology/horticulture and marketing.

The PSAB has a separate grower mandatory assessment for the purpose of supervising processor uniformity in grading and price posting of raw product. The

USDA administers grading of the final product, whether retail or industrial. This mandatory assessment has measurable performance outcomes, providing for better quality and increased demand for processed berries, thus justifying the assessment.

The continuous growth in acreage, production, and yield in the California strawberry industry has been a marketing challenge for the CSC, causing a gradual increase in the mandatory assessment to its maximum level of 5¢ per tray, resulting in a budget for selected years illustrated by the table below:

Table 31
CSC Budget (In Millions of Dollars)

Expenditures	1986	1993	1996	2000	2001	2004
Marketing	2.3	2.8	3.6	3.5	3.6	2.900
Consumer	1.3	.6	.8	.7	1.0	.063
Merchandising	.5	1.4	1.8	1.8	2.0	.907
Food Service	.5	.8	1.0	1.0	1.3	.449
Research	.5	1.0	1.5	1.5	1.8	1.000
Total	3.2	4.3	5.3	6.8	7.8	7.0

The 4 distinct phases of CSC programs to effect demand, CSCPROMO, reflected in these budgets, are Early, Media years, Incentive payments and the current Attitude and Behavioral phase, which uses proxies for sales as barometers of program effectiveness since media and retail incentive payments have been discarded. All of the budgets are not included in the previous table; however the Chapter 6, CSCPROMO variable discussion does include additional budget detail. The early years began with no advertising, little promotion, and no field representatives to make marketing presentations to retailers or food service operators. Public relations with recipes and press releases were the program of choice. During 1974–1990, the CSC programs used to influence the consumer, increase demand, price, and grower returns were TV, radio, tie-in ads, display contests, and public relations. Marketing budgets increased; however, by 1990 the media years ended because of cost and budget concentration was transferred to retail incentive payments and category management. Although the number of retail ads increased during this period, a new CSC direction was undertaken in 2003 as incentive payments seemed to only increase the retailer's bottom line, since it was believed that retailers would advertise without the minor incentive payments. This judgment proved to be true. In the current phase, *The Red Edge Program* renewed and enlarged the continuous former nutrition and health agenda with nutrition research seeking the magic bullet of a specific scientific strawberry discovery that would revolutionize medicine and increase the demand for strawberries. After discarding effective network pull TV because of expense, the CSC then dumped push programs (low weight TV, radio, billboards, and incentive payments), because they did not have significant measurable performance outcomes, namely an increase in prices and farm income (Table 27). The CSC substituted a combination of push, mostly pull techniques, that appeared to rely on barometers of measurable performance outcomes which are attitudinal and behavioral, and serve as proxies for sales. Unless the mandatory assessments have caused prices and farm income to increase, the justification for CSC and other marketing boards to utilize mostly public relations pull techniques is questionable at best. Growers' dissatisfaction with many state and federal marketing orders relates to the failure of prices and farm income to keep up with inflation, after paying mandatory assessments for programs designed to accomplish that goal. The inclusion of proxies as a non-price demand shift variable is also questioned in the economic literature (Lee), since it must be assumed that the proxies for sales are highly correlated with actual sales.

The private sector has the real burden of affecting the demand-shift variable, PSPROMO-CP. The vertically integrated grower/shippers have undertaken much of CSC category management function, as well as their normal sales, promotion, and distribution role. The retail advertising/promotion of strawberries is tending to become a partnership of private sector grower/shippers and retailers, all of whom are attempting to replace the traditional, fragmented, daily sales orientation of the fresh produce market with partnerships focusing on year end results (Cook). I have presented much detail, based on industry interviews as well as personal experience, on pre-commitment programs and contract pricing. The concept of sales planning or upward price management is a strategy between individual shippers and retail/wholesalers, which has reduced the percentage of spot market sales from a range of 50%–75% in the 1980s to 25% or less in the 1990–2000 period (Cook, Karst). An important element of PRPROMO and CP is their effect on the daily and weekly price level and price stability through the use of pre-committed and/or contract pricing. It is my view that these 2 strategies have reduced the extremely low prices that can linger for some period due to time lags between price and ad projections and commitments and the actual retail ad. Failure to project accurate weekly prices and volume in advance of an actual volume increase, results in prices below what the equilibrium price should have been with perfect knowledge. Under these frequent circumstances, because of imperfect volume projections, prices will not stabilize until shippers provide chains with prices and volumes for promotions that retailers implement with ads.

The crucial point is that this demand shift variable, (PSPROMO-CP), will cause demand to increase 2–3 times over the non-advertised product (Achabal). Table 8, illustrates the number of ads from 1991–2001. Advertising data from May 2003–2005 indicate a considerable increase from 1,377 to 1,544 ads during the week of May 20th alone, and without the CSC incentive program. Generally, a price decrease accompanies an ad offer, which may result in a retail price decrease, or it may not, depending upon the retailer willingness to reduce, maintain, or increase margins. The ultimate goal of a reduced shipper price, which reflects market conditions, is that it be passed on to consumers with lower retail margins. The real world is somewhere in between.

Q and PSPROMO-CP are the most important non-price demand shift variables in the model. CSC trade advertising and consumer public relations were important in earlier years of industry development when little product or nutrition information was available to retailers, consumers or the food service industry. The CSC has continued public relations activities, beginning in the second marketing phase, utilizing print, radio, and TV, to disseminate product information of all kinds, including nutrition, packaging, recipes, etc. Wearout may have already occurred and the industry's only barometer for measuring current or future CSC effort is the proxies for sales, assuming present CSC programs. Each Board must determine whether these proxies for sales represent sufficient measurable performance outcomes to justify mandatory assessments for the CSC public relations and nutrition programs.

Although I have included FVPI as a demand shift variable, because the demand for freezer product is an integral part of the total demand for strawberries, the specific demand shift effect of the CSC's programs is on the consumer, foodservice operator, industrial user, and export market. However, I have also included FVPI as a supply shift variable, but only insofar as freezer demand impacts the supplies available for the fresh market and price and not the total supply of strawberries. The relationship between the fresh and freezer markets has already been outlined in the section on Supply variables, emphasizing that the processor demand for freezer berries, affects the supply for the fresh market as well as fresh price. The reduction in US exports to Japan because of Chinese competition will potentially decrease the demand for freezer supplies and make additional supplies available for the fresh market, unless other domestic or foreign frozen users take the additional supply at the same or lower prices. If this does not occur, additional supplies would be available to the fresh market, reducing fresh market prices. Processing prices do place a floor under the fresh market, but the floor can vary downward, changing the dynamics of the market place. In this sense, fresh market volume is a function of the freezer markets. Even though it appears that Chinese expansion will continually impact US export to Japan, the CSC could at best, publicize the quality and health standards of the US frozen industry, and the necessity of multiple suppliers. An appraisal should be made of the marketing expenditures for processed strawberries in Japan and other far eastern markets for measurable performance outcomes from CSC export programs. Since exports are significantly declining with little hope of reversal, are proxies for sales sufficient to justify a mandatory assessment that includes funding for marketing processed strawberries? The processors have the responsibility and capability of marketing their product in the new competitive export environment, and the CSC would have a marginal effect at best. A clear delineation of cost would enable the CSC to reduce mandatory assessments by the amount of the reduced export marketing expenditure.

The final Chapter is on the Supply variable, S = f (Acreage, production, yields per acre, costs per acre, and net returns per acre), or S = (APYPR). The supply side of the model has been and is the main variable affecting price, volume, production costs, and grower returns. It is the most effected by CSC expenditures. The cooperation between the CSC and the University of California has provided the model for agricultural research in every aspect of the production of strawberries including nursery stock for commercial use. Voth and Bringhurst were interested and involved with nursery production of patented varieties because of the ability of a new variety to be a good nursery runner producer with early root development, for early varieties or short day types, as well as for everbearers. They were also concerned with the nursery cull rate because of the increased cost of production, which would then be passed on to the commercial grower. The new Ventana variety has nursery as well as commercial problems mentioned earlier. In Oxnard and all other southern California districts, earliness is the key to survival, and the Ventana variety is a poor early runner producer, meaning a significant nursery cull-rate and increased costs, resulting in much higher commercial grower costs. There appears to be insufficient university consideration to this important nursery aspect of the California strawberry industry, and is another failure to adequately cooperate with all industry members.

The uncontrollable seasonal characteristics of strawberries, including large volume increases and decreases, cause prices to rise and fall dramatically. The volume in the market explains a major portion of the level and variation in prices and demand is mainly a function of price. Therefore, the university CSC—programs in pomology and horticulture is vital for efficient production practices, which can augment yields and quality, at price levels related to total acreage and volume, and increase grower net farm income.

The research model, including pomology, horticulture, entomology, and plant pathology, the University's Foundation Plant Service (FPS) clean stock program, the University's Office of Technology Transfer Program (OTT) for licensing and patent; and the California Department of Food and Agriculture Strawberry Registration and Certification Program, has been and continues to be the foundation of increased yields, reduced or stable costs per acre, and improved quality. The data[5] [6] on grower production costs and average prices, indicates that there is little margin for grower cultural, harvest, or varietal error, since reduced yield per acre, inferior quality (reduced demand and price), and rising production costs, provide the conditions for the perfect storm for lower net farm income.

The current CSC Red Edge Program is funding research programs that have the potential for discovering the silver bullet, or a qualified health claim that goes beyond the limited claim that strawberries *may* improve heart health and memory and reduce the risk of some cancers. The qualified health claim that is sought would say that California strawberries, as part of a diet, have been shown to:

1. Reduce the risk of certain cancer.
2. Reduce the risk of heart disease.
3. Support cognitive function (memory).

The success of this CSC public relations program must then be measured as a proxy for sales and assessed as to whether there is a correlation with actual sales. If there is no silver bullet, the CSC will still have mostly proxy for sales barometers for measuring the effect of CSC programs. The question that remains for the CSC Board, or any commodity board, is whether the proxies for sales barometer provides sufficient justification for a CSC budget of $7–$8 million annually. To put it another way, what size budget for public relations is rational, with the knowledge that media advertising, incentive payments, and trade advertising (recently discontinued with a $45,000 budget) have been discarded?

My research revealed that an aerial survey, conducted by the University of California Office of Technology Transfer, provided information that there are

---

5   "Sample Costs to Produce Strawberries," by the University of California Cooperative Extension. (2004) Central Coast Section, South Coast Region (Oxnard): Yields and Returns, Table C, p.5; Net Returns Per Acre Above Total Cost, Table 4, p.17; South Coast Region (Santa Maria): Yields and Returns, Table C, p.6; Net Returns Per Acre Above Total Cost, Table 4, p.17Central Coast Region (Watsonville-Salinas): Yields and Returns, Table C, p.5; Net Returns Per Acre Above Total Cost, Table 4, p.15

6   "Sample Costs to Produce Strawberries," by the University of California Cooperative Extension. (2004) Central Coast Section, Table C, p. 5.

perhaps 2,000–3,000 commercial acres unreported for patent payments. This could possibly translate into approximately 8 million crates, or $400,000 in annual assessments, not received by the CSC. It would appear there is faulty enforcement of California nurseries, since *they* are the plant suppliers. I also wonder if, due to deficiencies in the university breeding program, the OTT is aware of the number of private plant breeding activities, as acreage is shifted from patented to proprietary varieties, thereby reducing university patent fees. The successful trend of the many private, non-university breeding programs effectively challenges the CSC-university propagation program and is indicative of the serious deficiency in the current pomology model. This has been accomplished without the advantage of utilizing university breeding stock other than released variety seed. Perhaps a new competitive model should be developed, which would include the CSC contracting with multiple plant breeders. The relatively short harvest season for Ventana, because of quality, appears to have caused the reemergence of the older variety Camarosa; a rare occurrence once a new variety has been introduced, further raising the question of the adequacy of the university breeding program. In an August issue of Newsweek, 2005, a Whole Foods Market produce expert commented that Camarosa has a good flavor and is deep red, without white around the stem. The new variety Albion on the other hand, appears to be an excellent day-neutral replacement for the discredited Diamante, and is illustrative of the need for multiple breeders, the university and competitors, with the growers ultimately deciding the victor in the marketplace. The animal spirits of the private sector may prevail. If the CSC and university do not address these trends, 60 years of cooperation could rapidly disintegrate. It has been a wonder to the California strawberry industry not only to witness this shift from the public UC patented strawberries towards proprietary varieties, but also to be capable of understanding why the relationship between the CSC and the UC Davis strawberry breeders has deteriorated from what was once a common and strong partnership, to becoming members of a common industry that now act like islands unto themselves and appear to be working only towards self serving goals. The deterioration of this past partnership relationship, between the CSC and the UC Davis strawberry breeders, has corresponded with the shift from the public UC strawberry cultivars towards the proprietary strawberry varieties. The California strawberry industry had been revolutionized a few decades ago by not only the introduction of superior public strawberry varieties, but also by horticultural and cultivation advances that propelled the strawberry fruit production levels per acre unit to totals that were once thought to be unobtainable levels. The research that was conducted at the two UC Davis field station sites, one in the Watsonville area and the other near Irvine, were used not only to introduce new promising strawberry cultivars, but also to advance the methods used by fruit growers to control a multitude of pests, salinity control measures, nutrient advances, horticultural advances with new drip irrigation systems, the use of various colored mulches to manipulate the fruit production cycle and a mass of other research topics. Now the research stations are restricted to most researchers, even other UC Davis researchers, and the work conducted at the field stations are now used basically for the advancement of new strawberry cultivars and not used for most of the many research projects that have been funded by the CSC. The reasons for this lack of team work appears to be a mystery to many of the researchers and strawberry growers and this lack of team effort has divided the research segment of the strawberry industry into separate and isolated cells.

Although the CSC is no longer engaged in generic advertising in the media directly, a message is promoted through press releases and other public relations activities. There are numerous cases against generic promotion campaigns, including challenges to the federal programs for honey, peaches, and nectarines, and several state programs, including grapes, cherries, raisins, milk, and cheese.

By leaving open the question of attribution in the 2005 Supreme Court beef marketing order decision, it must address this issue in each specific grower's case against generic advertising and assess the actual message control of government authorities to determine if free speech has been violated. The ruling stated that since the advertising is government speech, no speech violation occurred. There does not appear to be evidence that generic advertising campaigns have been shown to involve, or be influenced by, government claims to be the speaker. It may be that because of the 2005 Supreme Court decision on the beef industry, government will now become more involved in marketing order message management and/or that marketing order messages might now include a reference to a state or federal agency. This possible government intrusion in the functioning of heretofore independent growers' organizations would be an unwelcome, negative development for all marketing orders generally and agriculture in particular.

The one remarkable and glaring omission in the Supreme Court dialogue, and the growers' challenges to marketing orders, is the failure to address the importance and necessity of evaluating the commodity board's policies and programs and measuring its effectiveness. A sunset clause should be required of both State and Federal marketing orders providing for mandatory review of prior year activities, similar to a CPA's year-end audit by an outside organization or third party. Perhaps this would satisfy much of the current grower dissatisfaction and criticism.

My hope is that the framework for measuring program effectiveness, which this book has presented, will be of value to all marketing boards (See appendix) in becoming more accountable to participating growers.

Picture 30. Top: Mark Murai, CSC Interim President, 2005
Bottom: George Faxon, Manager, PSAB, and Roger Wyant, Assistant Manager

Picture 31. Dr. Royce Bringhurst and Victor Voth.
Top: Retirement Celebration in Monterey, after 40 years of service: 1950–1990
Bottom: Together again 10 years later

Biography

Born December 2, 1926 in Fort Wayne, Indiana, Herbert (Herb) Elliott Baum is the son, grandson, and great-grandson of wholesale produce merchants, who were the largest in northwestern Indiana. Baum's early involvement from the ages of 9 to 14 years in the family business founded by Samuel Baum, a Jewish Russian immigrant, ended as the Great Depression and family illness forced its closure. This experience would be the underpinning of his return to the produce business years later. Following his WWII service, he continued his education, MA '51 Economics, University of Chicago (alma mater of his great aunt and mentor, Minnette Baum: early Zionist, Suffragette and civil rights activist), and career as an agricultural economist in Washington, DC with the Office of Price Stabilization, Pan American Union and the US Department of State.

In 1953, Baum's interest and background in the private sector prompted the relocation to California and his first agribusiness employment at Blue Goose Inc., a nationwide grower, packer, and shipper of fruits and vegetables, marking his return to the produce business, however, this time in the largest agricultural area of the US. Having learned about growing, packing and shipping in all California growing areas, from balance sheets to horticulture and sales, and as assistant to the President, Baum's experience expanded into national fruit and vegetable sales until the relatively *new* strawberry industry was discovered by Blue Goose in 1955–56. The opportunity to become an entrepreneur in a thrilling new industry was reminiscent of his early years surrounded by a family of free market entrepreneurs. As the revolution in horticulture and pomology began, he became a part of that important history. In 1958 he joined Naturipe Berry Growers in San Jose as Vice President of Sales, the preeminent strawberry grower-shipper cooperative in California founded in 1917. The company, facing economic difficulty caused by early over production, increasing yields, and lagging marketing efforts in a revolutionary industry, was regenerated through Baum's successful efforts to expand the growing and marketing operations into all desirable California growing areas thereby creating an industry leader, which it remains today. After 33 years he retired as President in 1991.

As twice-elected chairman of the California Strawberry Commission, Baum was recognized as the industry spokesman for emphasizing advertising and marketing programs and firmly establishing the pomology and horticultural research model with the University of California at Davis, which he warns is in jeopardy. Although retired, he is an active industry student and in this book shares his experience, observations, and recommendations.

Herb and his wife, Gloria, enjoy hiking the red rock country of Sedona, AZ, cross-country skiing and white water rafting.

Appendix 1

This appendix indicates the scope of Federal and California state marketing orders. Many other states have similar marketing orders, as well.

Federal Marketing Orders

Fruit and vegetable industries voluntarily enter into marketing orders and marketing agreements which are designed to help stabilize market conditions and assist farmers in allowing them to collectively solve marketing problems with Federal oversight of certain aspects of their operations. The Marketing Order Administration Branch of the Fruit and Vegetable Programs oversees the programs to make sure the orders and agreements operate in the public interest and within legal bounds. There are currently 34 active marketing agreement and order programs, which collect assessment fees to cover operation and administrative costs of the programs.

A marketing order is a legal instrument authorized by the U.S. Congress through the Agricultural Marketing Agreement Act of 1937. Marketing orders differ from marketing agreements: marketing orders are binding on all individuals and businesses classified as "handlers" in a geographic area covered by the order while marketing agreements are binding only on handlers who are voluntary signatories of the agreement.[7]

---

[7]   USDA, Agricultural Marketing Service, Fruit & Vegetable Programs, Website (2005)

Federal Fruit & Vegetable Marketing Orders and Agreements:

A=Authorized but currently not used   E=In effect

State(s)/ Commodity(ies)	Research	Market Development/ Promotion	Committee Headquarters	Year Effective
Florida Citrus Fruit			Lakeland, Florida	1939
Texas Oranges & Grapefruit	E	E*	Mission, Texas	1960
Florida Avocados	E	A	Homestead, Florida	1954
California Nectarines	E	E*	Reedley, California	1958
California Pears & Peaches	E	E*	Reedley, California	1939
California Kiwifruit			El Dorado Hills, California	1984
Washington Apricots	A	A	Yakima, Washington	1957
Washington Cherries	A	A	Yakima, Washington	1957
Washington-Oregon Fresh Prunes	A	A	Yakima, Washington	1960
California Desert Grapes	E	A	Indio, California	1980
Oregon-Washington Winter Pears	E	E*	Milwaukie, Oregon	1939
Hawaii Papayas	Program	to be	Terminated	1971
10 states-Cranberries	E	E	Wareham, Massachusetts	1962
7 states-Tart Cherries	A	A	DeWitt, Michigan	1996
Washington-Oregon Bartlett Pears		E	Milwaukie, Oregon	1966
California Olives	E	E*	Fresno, California	1965
Idaho-Eastern Oregon Potatoes			Idaho Falls, Idaho	1941
Washington Potatoes			Moses Lake, Washington	1949
Oregon-California Potatoes	Program	Suspended	Indefinitely	1942
Colorado Potatoes	A	A	Monte Vista & Greeley, Colorado	1941
Virginia-No. Carolina Potatoes			Melfa, Virginia	1948
GA Vidalia Onions	E	E*	Vidalia, Georgia	1989
Walla Walla Onions	E	E*	Walla Walla, Washington	1995
Idaho-Oregon Onions	E	E*	Parma, Idaho	1957
South Texas Onions	E	E	Mission, Texas	1961
Florida Tomatoes	E	E*	Orlando, Florida	1955
South Texas Melons	E	E	Mission, Texas	1979
CA Almonds	E	E*	Modesto, California	1950

OR-Hazelnuts	E	E*	Portland, Oregon	1949
CA Pistachios			Fresno, California	2004
CA Walnuts	E	E	Sacramento, California	1948
Far West Spearmint Oil	E	E	Kennewick, Washington	1980
CA Dates	E	E*	Indio, California	1955
CA Raisins	E	E*	Fresno, California	1949
CA Dried Prunes	A	A	Sacramento, California	1949

Source: USDA, http://www.ams.usda.gov/fv/moabmotab.htm, Last Revised November 2, 2005

California State Marketing Orders

Similar to federal programs, state marketing programs are entirely self supporting. The industry pays operating costs, including the cost of government oversight. There are no subsidy payments to producers or handlers from general state tax sources and no general tax funds are used to support marketing program operations. Although only industry money is used to capitalize these programs, the taxing power of the state is used to collect these funds. State oversight is provided to assure that each marketing program operates in the public interest.[8]

There are 50 active state marketing orders, representing 40 agricultural commodities; of those programs, 25 are marketing orders and 2 are marketing agreements operating under the marketing act, while 20 are commissions, and 3 are councils operating under individual enabling legislature.

California Agricultural Marketing Orders:

Commodity	Headquarters
Alfalfa Seed Production Research Board	Dinuba
California Apple Commission	Fresno
California Artichoke Advisory Board	Castroville
California Asparagus Commission	Stockton
California Avocado Commission	Irvine
California Beef Council	Sacramento
Buy California Marketing Agreement	Sacramento
Cantaloupe Advisory Board	Dinuba
California Fresh Carrot Advisory Board	Dinuba
California Celery Research Advisory Board	Dinuba
California Cherry Marketing Program	Lodi

[8]   California Department of Food & Agriculture, Marketing Branch Website (2005)

Citrus Research Board	Visalia
CA Citrus Nursery Research & Education Program	Sacramento
California Cut Flower Commission	Watsonville
Dairy Council of California	Sacramento
California Date Commission	Indio
California Fig Advisory Board	Fresno
Dry Bean Advisory Board	Dinuba
California Milk Processor Advisory Board	Berkeley
California Forest Products Commission	Auburn
CA Garlic & Onion Dehydrator Advisory Bd	Stockton
California Garlic and Onion Research Board	Clovis
CA Grape Rootstock Improvement Commission	Sacramento
California Kiwifruit Commission	El Dorado Hills
Lake County Winegrape Commission	Lakeport
California Lettuce Research Program	Salinas
Lodi-Woodbridge Winegrape Commission	Lodi
CA Manufacturing Milk Advisory Board	Modesto
California Milk Producers Advisory Board	South San Francisco
California Melon Research Board	Dinuba
California Cling Peach Growers Advisory Bd	Dinuba
California Pear Advisory Board	Sacramento
California Pepper Commission	Dinuba
California Pistachio Marketing Agreement	Fresno
California Pistachio Commission	Fresno
California Plum Marketing Program	Reedley
California Potato Research Advisory Board	Dinuba
California Dried Plum Board	Sacramento
California Raisin Marketing Board	Fresno
Rice Research Advisory Board	Yuba City
California Rice Commission	Sacramento
California Salmon Council	Folsom
California Sea Urchin Commission	Sacramento
California Sheep Commission	Folsom
California Strawberry Commission	Watsonville
Processing Strawberry Advisory Board	Watsonville
California Table Grape Commission	Fresno
Processing Tomato Advisory Board	Davis
California Tomato Commission	Fresno
California Walnut Commission	Sacramento
California Wheat Commission	Woodland
California Wild Rice Board	Sacramento
Winegrape Inspection Marketing Program	Dinuba

Source: CDFA, Marketing Branch, http://www.cdfa.ca.gov/mkt/mkt/mktbrds.html

Appendix 2

The UC Davis study[9], May 2004, which discussed the USDA National Berry Report and the relationship between the spot market and precommitted sales indicated that the "directly observed effect of market precommitments is to lower prices relative to spot market prices in a given week." The conclusion of the analysis did not appear to take into account that the eventual price, unknown by shippers at the time and consequently by the USDA, of the unsold product on any given day and week will probably be lower than the spot market; for that reason the precommitted daily or weekly prices that may appear lower (USDA data) could actually be higher than the "real" spot market price. The study did indicate, however, that "additional analysis provided some evidence that the precommitments may increase shipper returns by increasing total strawberry demand." The latter statement is valid as proved by the widely used practice of advance price and volume precommitments and contract pricing. The practice of sales planning, or upside price management, is based upon supply and price projections that are attempts at prevention of severe price declines resulting from underestimating supply and failure to price and promote for the future supply level. The industry knows with certainty that promotion increases demand although the study states that "the demand increase *may* be fueled by retailer promotions, which are commonly supported by precommitments."

I suggest that the USDA National Berry Report is interesting and provides some guidelines on the daily market, but is of little if any value in facilitating the analysis or providing useful or accurate information on the nature or relationship of the spot market to either precommitments or contract pricing.

---

[9] Carter C., J. Chalfant,R. E. Goodhue, and S. Mohapatra, *Private Sales and Public Information: Does the USDA's National Berry Report Provide Information on the Relationship between the Spot Market and Precommitted Sales?* Agricultural Marketing Resource Center, Department of Agricultural and Resource Economics University of California, Davis. May 2004

Glossary of Terms

Abiotic factors are those non-living physical and chemical factors which affect the ability of organisms to survive and reproduce.

Asexual propagation involves the vegetative parts of a plant including roots, stems, or leaves. A part of a single parent plant is made to regenerate itself into a new plant, which is genetically identical to the parent plant. The strawberry plant regenerates itself through runner production, which enables much faster and exact replication of the parent.

Barometers are indicators or measurements of particular CSC policies.

Big box store is a retailer, such as Wal-Mart, Costco, Sam's Club, or Target, which offers an enormous variety of merchandises in a massive free-standing store.

Biotic factors are all the living things or their materials that directly or indirectly affect an organism in its environment. This would include organisms, their presence, parts, interaction, and wastes. Factors such as parasitism, disease, and predation (one animal eating another) would also be classified as biotic factors.

Brix, or sugar percentage of fruit which ensures proper sweetness and maturity, involves a system for measuring the sugar content of a solution at a given temperature using a hydrometer for measuring.

Cat-facing is the unsightly dimpling, deformity and scarring of fruit resembling the puckered cheeks of a cat, and is caused by a number of insects, mainly the Lygus bug, with a piercing-sucking feeding habit.

Catching down runners is a substitute for planting nursery stock and produces lower yields of smaller berries, which have a larger proportion of misshapen berries and higher incidence of disease.

Check-off programs refer to the generic research and commodity promotion programs for farm products that are financed by assessments applied to sales of those products by producers, importers, or others in the industry.

Chloropicrin is a preplant soil fumigant. Like most poisonous fumigants, chloropicrin is a Restricted Use Pesticide so its distribution and use are highly controlled. Since it does not have the excellent herbicidal properties of methyl bromide or the broad nematocidal properties of 1, 3-D, chloropicrin's use as an alternative will be in conjunction with 1, 3-D and other compounds with broader herbicidal properties. See also Methyl bromide and Soil Fumigation.

Clone is a progeny of an individual strawberry plant which was produced asexually

Cultivar is a variety of a cultured plant that is developed by breeding and has a designated name.

Cyclospora cayetanensis are microscopic, one-celled organisms that can contaminate fresh produce and burrow in the small intestine causing diarrhea like symptoms. The illness can be treated with antibiotics or could pass naturally within a period of a few days up to a month. The washing of fresh produce may prevent some food-borne illnesses, but may not prevent cyclospora infection.

DNA Fingerprinting is a method of identification that compares fragments of deoxyribonucleic acid (DNA); it is sometimes called DNA typing. DNA is the genetic material found within the cell nuclei of all living things.

Day neutral, or Everbearer, refers to a cultivar, which is continuously producing fruit from May until it rains sufficiently to stop commercial production in late fall. The main difference between day neutrals, developed in California, and everbearers, developed in the northeastern part of the United States, is that day neutrals have a lower temperature threshold of growth and continue to grow in southern California and produce fruit going into the winter months and continuing. The everbearer developed in the northeastern states will not vegetate during this period or produce fruit continually until the temperature becomes warmer. The words tend to be interchangeable, with University of California cultivars considered day neutrals and some proprietary varieties, everbearers. Both are insensitive to day length and produce buds, fruit and runners continuously if temperatures remain above 35° F and below 85° F. Some examples of day neutrals are the university patented varieties Irvine, Selva, Diamante, Albion and proprietary varieties of Dricoll, Wellpict, and Nelson.

Demand shift variables can be effected by outside factors such as policies introduced by the CSC. Examples would be advertising, promotion and changes in behavior and attitudes.

Elasticity of demand (with respect to price) is a situation in which price elasticity of demand exceeds 1 in value. This means that the response of buyers to a price reduction is sufficiently large that total revenue (price time quantity purchased) rises. Conversely, if price were to be increased, total revenue would go down.

Elasticity denotes the responsiveness of one variable to changes in another. The elasticity of x with respect to y means the percentage change in x for every 1 percent change in y.

Everbearer: See Day neutral

Explant is a plant part taken from a living plant and placed in a culture medium.

Foundation stock means strawberry plants that are first year propagation from plants that have been approved on the basis of annual indexing.

Frigo plants are dug in January and stored at 27°F–30°F.

Fumigation, See Methyl bromide and Soil Fumigation.

Generic advertising is the industry's media advertising of a particular commodity without reference to a specific producer, brand name, or manufacturer funded through assessments called *check-off programs.*

Generic promotion involves generic instore promotions, couponing, special allowances, and incentive ads for a commodity funded through assessments called check-off programs (See Push Marketing).

Gross Rating Point (GPR) is a unit of measure in advertising audience size that represents the total delivery or weight of a media schedule during a specified time period. GRPs are calculated by multiplying the reach of the media schedule by the average frequency.

High Gallonage Sprayer distributes more water per acre mixed with chemical materials which prevents spray burn on new flowers, leaves, and fruit, if present, avoiding cat-facing and other fruit deformities and defects. This is preferable to the low gallonage application, which provides a higher distribution of concentrated chemical materials which may burn the plant, causing the defects listed above. See illustration, Picture 28 on page 138.

Horticulture is the science and art of growing fruit, vegetables, flowers, shrubs, and trees.

Index means to test for virus infection by making a graft with tissue from the plant to be tested to an indicator plant or by other methods approved by the director of the California Department of Food and Agriculture's Strawberry Registration and Certification Program.

Inelasticity of demand (with respect to price) is the situation in which price elasticity of demand is below 1 in value. This means that when price is reduced, total revenue (price times quantity purchased) goes down and when price is increased, total revenue goes up. Perfectly inelastic demand means that there is no change at all in quantity purchased when prices go up or down.

Isoenzyme is one of a group of related enzymes catalyzing the same reaction but having different molecular structures and characterized by varying physical, biochemical and immunological properties.

Isoenzyme analysis technology (IEF) is one of various methods for the identification and authentication of cell lines recently replaced by cytogenetic analysis and DNA fingerprinting.

Locus, a gene or genetic marker

Measurable performance outcomes refer to the direct effects of CSC policies on demand, prices, crop yields, quality, and grower's net farm income. *Proxies* may indirectly affect demand by influencing behavior and attitudes, which are difficult to measure.

Meristem is a process of taking the smallest growing point (*meristem tip*) from the plant and growing it in a medium within a test tube; when the plant is mature enough, it is transplanted to begin propagation.

Meristem tip is an *explant* comprising the meristem (meristematic dome) and usually one pair of leaf primordia. Do not confuse the meristem tip with the term "shoot tip," which is much larger and usually has more immature leaves and stem tissue.

Methyl bromide (MeBr) is an odorless, colorless poisonous gas that has been used as an agricultural soil and structural fumigant to control a wide variety of pests. However, because MeBr depletes the stratospheric ozone layer and is classified as a Class I ozone-depleting substance, the amount of MeBr produced and imported in the U.S. was incrementally reduced until the phaseout took effect on January 1, 2005, except for allowable exemptions. These exemptions include the Quarantine and Preshipment (QPS) exemption, to eliminate quarantine pests, and the Critical Use Exemption (CUE), designed for agricultural users with no technically or economically feasible alternatives. See also Chloropicrin and Soil Fumigation.

Model is a system used to analyze the relationship between the quantities demanded and the variables affecting demand, including prices, inventories, prices of substitutes, and total income for all goods.

Nominal Prices relate to current prices considered in terms of the stated or original value only, ignoring changes due to inflation and other factors.

Nuclear stock means strawberry plants, which were originally indexed, and their progeny have been regularly reindexed and protected continuously from virus infection by federal and state entities.

Nuclear meristem stock are strawberry plants (propagated with meristematic tissue from a plant which has had heat therapy) that have been originally indexed and found free of known viruses by federal and state agencies.

Phytotoxicity means being poisonous to plants.

Plant propagation is the process of multiplying the numbers of a plant, perpetuating a species or maintaining the youthfulness of a plant. There are 2 types of propagation, asexual and sexual.

Pomology is the study or practice of cultivating fruit.

Point of Purchase Material (POP) is a technique which offers the trade attractive materials for in-store use in building displays and attracting consumer attention within the store.

Precommitment is an informal, usually verbal agreement between a shipper and retailer whereby a portion of the crop is promised for sale at a price agreeable to both parties.

Price elasticity of demand is a measure of the degree to which quantity demanded by buyers responds to a price change. The elasticity coefficient, or quantitative measure of elasticity, is: percentage change in quantity bought divided by percentage change in price.

Proprietary Varieties are developed independently by private firms and are patented for sale or private use by growers, shippers, and processors.

Proxies for sales, or substitutes are used instead of measuring actual sales, for example product usage, consumer attitudes, shopping behavior and consumer perceptions of the product. Proxies are considered as actual figures relating to sales when actual sales measurement are not possible. However, in order to consider a *proxy* as an actual sale, it must be assumed that it is highly correlated with actual sales.

Pull marketing includes consumer advertising, public relations, consumer's sweepstakes or contests and couponing; techniques all designed to directly influence the consumer.

Push marketing is the traditional commodity board approach and is designed to directly encourage retailers or food service users to feature and display a give commodity because their sales and profits would increase.

Raw Weighted Value (RWV) is the accumulative strength and frequency of advertising based on the size of market and retailer.

Real price is the nominal price after changes in inflation have been taken into account.

Sales Planning, See Upside price management.

Salvage market is a depressed market for strawberries that cannot be harvested and sold at current fresh market prices because of quality.

Sexual propagation involves the union of the pollen from the male with the eggs from the female in order to produce a seed which develops into a primary root and plant

Short day type refers to a cultivar that bears fruit early in the year because it has the ability to produce fruit during short daylight hours. Examples are Chandler, Camarosa, and Ventana.

Soil Fumigation is a method of pest control whereby an area of land, which has been carefully covered and sealed with plastic, is injected with poisonous, gaseous pesticides to suffocate or poison the pests, diseases, and weeds within. See also Methyl bromide.

Spot market is a market with immediately available goods; a market in which commodities, securities, or currencies are traded for immediate payment and delivery.

Tangible asset is represented by the barometers, which measure the effects of the demand shift variables on price, farm income quality, attitudes, etc, and is the major effect of policies.

Tectrol©, introduced in 1968 by Transfresh Corp., is a product sealed around each pallet of strawberries just before loading on refrigerated trucks. Once the strawberries are sealed, the degradation process is slowed and any Botrytus molds are inhibited.

Upside price management is a sales planning concept requiring each shipper to analyze historical industry demand, volume, and price data, in conjunction with the CSC, and then to project proprietary volumes and prices as well as industry volumes directly to retailers in the form of precommitments or contract pricing.

Wearout is the occurrence of consumers becoming so used to an ad that they stop paying attention to it.

References

Alston, J.M., J.A. Chalfant, J.E Christian, E. Meng, and N E. Piggott. *The California Grape Commission's Promotion Program an Evaluation*. Giannini Foundation Monograph Number 43. 1997.

Alston J. M., H. Carman, J. Chalfant, J. Crespi, R. Sexton, and R. Venner. *The California Prune Board's Promotion Program, An Evaluation*. Giannini Foundation Research Report Number 344: 42. March 1998.

Bayh-Dole Act. A Guide to the Law and Implementing Regulations, Council on Governmental Relations, September 1999.

Bertelson, Diane. *The US Strawberry Industry*, USDA, Economic Research Service, Statistical Bulletin No. 914, January 1995.

Blaylock, J.R. "Discussion: Ongoing Empirical Research on Generic Advertising." *Commodity Advertising and Promotion*. Ames: Iowa State University Press, 1992: 70–78, citing "Commodity Advertising: Theoretical Issues Relating to Generic and Brand Promotions." *Agribusiness* 1:269–276.

Bolda, Mark P., Laura J. Tourte, Karen M. Klonsky, Richard L. Loura. *Sample Costs to Produce Strawberries*. University of California Cooperative Extension. 2004.

Botsford, Ketchum Advertising Agency. *California Strawberry Advisory Board (CSC) 1977 Marketing Plan*, November 18, 1976.

Brader, Charles, K. Kesecker, and H. Ricker. *Government Policy and Program Information Needs. Commodity Advertising and Promotion*. Ames: Iowa State University Press, 1992: 343–354.

Browne, Greg and Ravindra Bhat, "Managing Disease Caused by Phytophthora on California Strawberries." *California Strawberry Commission Annual Production Research Report, 2003–2004 Research Projects*: 69–87.

California Department of Food & Agriculture. Strawberry Registration and Certification Program Regulations, Article 9. Regulations for California Certified Strawberry Plants. Section 3049

California Food and Agricultural Code, Division 22, Part 1, Section 63901, Part (b), 1995.31

California Food and Agricultural Code, Div 22, Ch 17.5, § 77401-77501, April 18, 1996.

California Strawberry Commission. Annual Reports, 1970–2001.

California Strawberry Commission. Annual Production Research Report, 2003–2004.

California Strawberry Commission. California Strawberry Export Report, 2005.

California Strawberry Commission. Category Management Report, 2001–2004.

California Strawberry Commission. Data Reports, 2000–2001.

California Strawberry Commission. Quarterly Reports, 2003–2005.

California Strawberry Commission. Peak Season Pricing Analysis Final Results, January 29, 2001 (From Perishables Group).

California Strawberry Commission. Radio Campaign Market Results, August–September, 2000.

California Strawberry Commission. Pricing Study Results, 1998–2000.

Carriere, Michael. *UC Technology Transfer Program: Mission Statement*, University of California Office of Technology Transfer, Oakland, California.

Carriere, Michael. *UC Technology Transfer Program: Process for Obtaining UC Patented Strawberry Cultivar Material*, University of California Office of Technology Transfer, Oakland, California.

Carter, C. "Economic Analysis Of The California Strawberry Industry: Price Determination and Marketing Practices." *The Pink Sheet*, California Strawberry Commission, February 15, 2001.

Carter, C., J. Chalfant, R. Goodhue, Zhang. "How Large is China's Strawberry Industry?" *The Pink Sheet*, California Strawberry Commission. June, 2005.

Carter C., J. Chalfant ,R. E. Goodhue, and S. Mohapatra, *Private Sales and Public Information: Does the USDA's National Berry Report Provide Information on the Relationship between the Spot Market and Precommitted*

Sales? Agricultural Marketing Resource Center, Department of Agricultural and Resource Economics University of California, Davis. May 2004

Carter, C., J. Chalfant, R. Goodhue, F. Han and Tian Xia. "Economic Analysis of the California Strawberry Industry: Price Determination and Marketing Practices." *The Pink Sheet*, California Strawberry Commission. February 15, 2001.

Carter, C., J. Chalfant, M. De Santis, R. Goodhue, and F. Han. *Economic Impacts of the Methyl Bromide Ban On The California Strawberry Industry: A Market Level Analysis.* Department of Agricultural and Resource Economics, University of California, Davis, July 2000, Revised January 16, 2001.

Cook, R. *Fresh Fruit and Vegetable Markets: Retail Structure, Trade Practices, and Pricing Strategies.* USDA. Cited by Produce Marketing Association Summit 2000 Workshop.

Darrow, George M., *The Strawberry: History, Breeding and Physiology.* New York: Holt, Rinehart and Winston, 1966: 227.

Duniway, John M. "Evaluation of Some Chemical and Nonchemical Alternatives to Methyl Bromide Fumigation of Soil for Strawberry." *California Strawberry Commission Annual Production Research Report, 2003–2004 Research Projects*: 88–103

Elrick and Lavidge, Inc. "Consumers Use and Opinions of Strawberries and Effectiveness of TV Ad Campaign." (unpublished Report for the California Strawberry Advisory Board, June 27, 1974)

Fennimore, Steven A. and Husein Ajwa, "Weed Management in Strawberry." *California Strawberry Commission Annual Production Research Report, 2003–2004 Research Projects*: 146–154.

Gubler, Doug, "Results of Field Trials for Fungicide Efficacy on Strawberry Powdery Mildew, Botrytis, and Anthracnose Fruit Rot." *California Strawberry Commission Annual Production Research Report, 2003–2004 Research Projects*: 114–121.

Han, Frank. "Explaining Seasonality in California Strawberry Prices." California Strawberry Commission. [2000?]

Han, Frank, C. Carter, and R. Goodhue. "Seasonal Prices and Supply-Side Adjustments in the California Strawberry Industry." *The Pink Sheet*. No. 99–6. California Strawberry Commission, April 6, 1999.

Hayes, Dermot. "Incorporating Advertising into Demand Systems." *Commodity Advertising and Promotion.* Ames: Iowa State University Press, 1992: 181–189.

Industry Interviews: March, 2002–October, 2005.

Johnson, S.R. "Future Directions for Advertising Research: A Researcher's Perspective." *Commodity Advertising and Promotion.* Ames: Iowa State University Press, 1992: 363–369.

Karst, T. "Contract Pricing Making Inroads." *The Packer*, December 6, 2004.

Karst, T. "Generic Promotion Fights Not Over." *The Packer*, May 30, 2005.

Karst, T. "Ruling Bolsters Government-Run Ag Checkoffs." *The Packer*, May 30, 2005.

Kinnucan, Henry, H Xiao, C Hsia, and J. Jackson. "Effects of Health and Generic Advertising On U.S. Meat Demand." *American Journal of Agricultural. Economics.* No.79, February 1997: 13–23.

"Marketing Order for Processing Strawberries, As Amended," State of California Department of Food and Agriculture, Marketing Branch, Effective August 23, 1967, incorporating amendments through March 12, 2001.

Krieger, Robert I., "Measuring and Mitigating Pesticide Exposure of Handlers and Strawberry Harvesters." *California Strawberry Commission Annual Production Research Report, 2003–2004 Research Projects*: 155–163.

Lee, Jong-Ying and Mark G. Brown. "Commodity versus Brand Advertising: A Case Study of the Florida Orange Juice Industry." *Commodity Advertising and Promotion*, Iowa State University Press: Ames, 1992: 206-221.

Martin, Frank N., "Development of Controlled Procedures for Black Root Rot Pathogens: Assessment of Host Tolerance, Cultivar Yield Potential When Grown in Nonfumigated Soil, Alternative Fumigants and Fungicides." *California Strawberry Commission Annual Production Research Report, 2003–2004 Research Projects*: 122–135.

McClelland, E.L., L. Polopolus, and L. Meyers. "Optimal Allocation of Generic Advertising Budgets." *American Journal of Agricultural. Economics.* v53, n4 (November 1971): 565–572.

McClure, Bob. "Citrus May See Future in Ruling." *The Packer*, May 30, 2005.

Mitchell, Gordon, E. Mitcham, J. Thompson, N. Welch. *Handling Strawberries for Fresh Market.* University of California, Division of Agriculture and Natural Resources, Publication 2442, 1996.

The Naturipe Story, Naturipe Berry Growers, 1967.

Perosio, Debra J., Edward W. McLaughlin, Sandra Cuellar, Kristen Park. "Supply Chain Management in the Produce Industry." Food Industry Management Program, Cornell University, 2001.

Rauser, G.C., E. Hochman. *Dynamic Agricultural Systems: Economic Prediction and Control.* North Holland Press, 1979: 15–21.

Richards, T.J., Patterson, P.M., *Retail Contracting in Fresh Fruit*, Morrison School of Agribusiness & Resource Management, Arizona State University. May 2002.

Roselle, Tracy. "Citrus Department Axes 12.2 Million From Budgets." *The Packer.* June 25, 2001.

Shaw, Douglas V. and Kirk D. Larson, "Strawberry Genetics, Breeding, Physiology and Production Management" *California Strawberry Commission Annual Production Research Report, 2003–2004 Research Projects*: 136–145.

Tomek, William G. & Harry M. Kaiser, "On Improving Econometric Analyses of Generic Advertising Impacts." Cornell University, from "Symposium on Improving the Science of Promotion Evaluation." March 19, 1999, Washington DC.

Toscano, Nick C. "Insecticide Efficacy Monitoring in Greenhouse Whitefly Populations." *California Strawberry Commission Annual Production Research Report, 2003–2004 Research Projects*: 20–29.

University of California Material &Custom Services. Foundation Plant Services (FPS), College of Agricultural & Environmental Sciences at the University of California, Davis. Website: http://www.ucop.edu/ott/strawberry/Customservices.htm

Voth, Victor. "A History of Strawberry Pomology and Horticulture: 1945-1990." (Unpublished Paper, 1991).

Ward, Ronald W., "Discussion: Generic and Brand Advertising." *Commodity Advertising and Promotion.* Ames: Iowa State University Press, 1992: 222–231.

Wells, George S., *Garden in the West—A Dramatic Account of Science and Agriculture*. Dodd & Mead & Co., 1970: 107–126.

Wilhelm, Stephen and J.E. Sagen., *A History of the Strawberry from Ancient Times to Modern Markets*. UC Berkeley, 1974.

Zalom, Frank, "Statewide Strawberry Entomology Research Program." *California Strawberry Commission Annual Production Research Report, 2003–2004 Research Projects*: 30–61

Index

978-0-595-37708-4
0-595-37708-4

www.ingramcontent.com/pod-product-compliance
Lightning Source LLC
Chambersburg PA
CBHW030920180526
45163CB00002B/406